# Practical Applications of Approximate Equations in Finance and Economics

# Practical Applications of Approximate Equations in Finance and Economics

### Manuel Tarrazo

QUORUM BOOKS
Westport, Connecticut • London

**Library of Congress Cataloging-in-Publication Data**

Tarrazo, Manuel, 1958–
    Practical applications of approximate equations in finance and economics / Manuel Tarrazo.
      p. cm.
    Includes bibliographical references and index.
    ISBN 1–56720–393–0 (alk. paper)
    1. Finance—Mathematical models.    2. Economics—Mathematical models.    I. Title.
HG106.T37   2001
330'.01'51—dc21        00–037266

British Library Cataloguing in Publication Data is available.

First published in 2001

Quorum Books, 88 Post Road West, Westport, CT 06881
An imprint of Greenwood Publishing Group, Inc.
www.quorumbooks.com

Printed in the United States of America

JK

**Copyright Acknowledgments**

The author and publisher gratefully acknowledge permission to quote from the following:

Caws, Peter. *Structuralism for the Human Sciences*. Humanities Press, 1988, reprinted 1997. Used with permission from Humanities Press.

Every reasonable effort has been made to trace the owners of copyright materials in this book, but in some instances this has proven impossible. The author and publisher will be glad to receive information leading to more complete acknowledgments in subsequent printings of the book, and in the meantime extend their apologies for any omissions.

*"For the nature of the scientific man contains a real paradox: he behaves like the proudest idler of fortune, as though existence were not a dreadful and questionable thing but a firm possession guaranteed to last for ever. He seems permitted to squander life on questions whose answer could be at bottom of consequence only to someone assured of an eternity."*

—**Fiedrich Nietzsche**
*Untimely Meditations*

*" . . . but the scientist still thinks there is something valuable to be found, and tries to catch and save that playful, afternoon sunray of knowledge perhaps to warm the evening of his life."*

—**M. T.**

# CONTENTS

# PREFACE

This is a monograph about practical applications of a class of models, approximate equations systems, to economics and finance. Approximate equations systems preserve the rigor of abstract, academic models while providing flexibility for practical decision-making. This monograph has been written for academic researchers and practitioners in economics and finance. However, its treatment of scientific modeling and methodology may also be of interest to researchers in applied mathematics, computer, or cognitive sciences.

The French philosopher Michel Foucault wrote the following in a thought-provoking book on the epistemology of human sciences:

> The fundamental codes of a culture—those governing its language, its schemas of perception, its exchanges, its techniques, its values, the hierarchy of its practices—establish for every man, from the very first, the empirical orders with which he will be dealing and within which he will be at home. At the other extremity of thought, there are the scientific theories or the philosophical investigations which explain why order exists in general, what universal laws it obeys, what principle can account for it, and why this particular order has been established and not some other. But between these two regions, so distant from one another, lies a domain which, even though its role is mainly an intermediary one, is nonetheless fundamental: it is more confused, more obscure and probably less easy to analyze. (1994, xx)

Foucault, among other issues, is addressing the uncomfortable void between academia and the common person, and likewise that between scholars and practitioners. From where does this divide originate?

Both scholars and practitioners face the same reality and strive to solve practical problems. In their pursuits, they both use the same tools: models. These are representations of reality, more or less stylized, more or less accurate, all of them problem driven.

Models generally consist of three elements: (1) a set of mathematical objects with their own rules of operation, (2) theories that describe the model within a context of reference, and (3) data that flow through the model and distill into critical information (intelligence) not previously available.

The practitioner's representation is often implicit, hard to explain, and driven by intuition, emotion, and remnants of theories, all of which are present when a decision must be made. Imagine, then, the frustration of a portfolio or financial manager when things do not go as planned. He faces requests for explanations of what happened and has no answers. Even though it is clear that something went wrong, we would probably decide that knowledge must have been insufficient at the time the decision was made. In that case, we might then examine the representations used or the decision-maker himself.

Scholars' models, derived from their own representations, are usually quite impressive. So impressive, in fact, they become "privileged" in the sense of both appearing in textbooks and by having to be used by anyone claiming to be an expert. Many factors are included and uncertainty accounted for by a veil of probability, which brings with it an army of well-behaved variables and assumed processes. Unfortunately it seems, in areas such as economics and finance, monolithic and self-contained models like these become unrealistic and hard to apply.

We need practical models that can adapt to the user's situation. Models in which the abstract—general, theoretical, and rational—leads to the specific of the situation and the subjectivity of the decision-maker. Practical also means intuitive. When we make a rational (abstract) model intuitive, we make it practical. Such a model would allow the practitioner's feel for things to come into play.

The German philosopher Schopenhauer made the same observations about 175 years ago, with similar wording. In his work, aptly entitled "The World as Will and as a Representation," subjective epistemology is the link between the idealized representation and reality itself. The same line of thought drives modern research in cognitive science: at the computing level—when decisions are made—the worlds of reality and speculation meet (Posner, 1993, several readings, especially pages 1–47).

What academic models need is built-in room for maneuvering. This, in fact, is precisely what the scholar has enslaved herself to avoid. In this monograph I present such a class of models: approximate equations systems (AES). AES are simultaneous equations systems (SES) with an attached error coefficient. The variation that coefficients may assume before the equations system breaks down can be an indicator of informational strength in the model. This error adds flexibility to the original SES, while preserving the delicate and comprehensive informational structure. Presence of error creates room for uncertainty and imprecision of types more varied and general than those allowed by standard probability theory alone.

Over time, significant effort has been put into developing better theories, better mathematical objects, and into defining and obtaining better data.

Surprisingly, comparatively less effort has been put into determining the accuracy of exact mathematical methods or questioning our basic presuppositions of model building.

Sometimes, our ignorance is covered up with language. For example, data on X and Y should not go together, but we change X into Z so that running a regression of Z on Y makes sense. In other cases, we disguise our ignorance by using probability theory as a veil over our exact model. This relieves us from changing the model, thus explaining the popularity of probabilistic methods. Probability theory turns what we do not know into an epiphenomenon, something inessential. In some cases, for example, option valuation, the size of the variance determines the value of a dependent variable without us having to explain what the determinants of that variance are. We further assume that the phenomenon studied exhibits ergodic regularity, if not determinism, and the uncertainty we choose to deal with becomes very manageable.

In humanistic, dynamic, and complex systems such as those in economics and finance there is, invariably, much more than probabilistic uncertainty. This "much more" calls for approximate methods such as those offered by AES. We will make our case for AES using practical examples from macroeconomic modeling, financial planning, and portfolio management. These are three very disparate examples and the state of knowledge in each case varies considerably. With what we do know about macroeconomic theory, it is hardly possible to build a practical and realistic macroeconomic model, even with the help of approximate equations. However, our knowledge about firms does make it possible to develop practical financial planning models with approximate equations. In the portfolio case, the theory is well known and approximate equations make it more attractive as an investment tool, but we still know little regarding the theory's effectiveness in creating wealth for the user.

AES are not the only way to handle the complexity of economic and financial systems, but they seem one of the most expeditious ways to start moving in the right direction. In this monograph, we will see how AES bridge the gap between abstract-rational (scholars'), and specific-intuitive (practitioners') representations, thus wedding theory to practice. The organization reflects its dual readership of practitioners and scholars, perhaps beyond the fields of economics and finance. Our emphasis is on clarity and making our models transparent enough so they become useful for practical purposes. Consequently, in spite of the necessary analytical sophistication required to deal with AES, we do not strive for symbolic wizardry or mathematical prowess. The focus is on developing models that address practical needs. Remember, we are trying to cast light on the "more confused, more obscure, and probably less easy to analyze" area mentioned by Foucault.

This monograph has three major divisions: (1) approximate equations, (2) applications, and (3) extensions. The first part introduces the reader to modeling (Chapter 1), simultaneous equations, and approximate equations (Chapter 2), where we provide our technique to solve these systems. The second part provides examples of AES applied to macroeconomic modeling (Chapter 3), financial planning (Chapter 4), and portfolio management (Chapter 5). In each of these applications, it is necessary to determine what ails each of these representations and thoroughly examine the implications of stating the model in its approximate form. The third part of the monograph examines the implications of AES for decision analysis (first part of Chapter 6), and modeling alternatives, such as fuzzy sets and neural networks (second part of Chapter 6). This final chapter is not exhaustive but simply outlines the relationships between AES and other modeling alternatives.

As noted earlier, this monograph stresses methodological issues. In my opinion, the most radical development in a student, scholar, or professional occurs when we realize the fundamental features and implications of what we handle (objects, theories, and data). This we very rarely do, and yet, this is "where the money is" to use a business expression. The stress on foundations also lays a bridge between our disciplines—economics and finance—and applied mathematics, computer science, cognitive science, and perhaps biology and psychology. This bridge is important because one may have the perception that any significant problem requires in-depth knowledge of the world both inside and out, and therefore, an interdisciplinary approach.

Finally, I also stress methodological issues because I believe that much of the apparent precision in complicated, multi-equation models is obtained at the cost of realism. It is only when we explicitly recognize the limitations of our methods that we begin to overcome them. This is what AES seem to offer. AES are not the final answer to our modeling problems, but only a step in the right direction.

## ACKNOWLEDGMENTS

This book would probably have not existed without the help of James McCann, whom I had the good fortune to meet when he was a McLaren Research Fellow and who has thoroughly and patiently proofread and edited the first drafts of this manuscript. With his help, I undertook the task of writing this monograph without the feeling of doing it alone. I also want to acknowledge the support and cheer received from my colleagues and deans at the McLaren School of Business. Thanks to Eric Valentine, from Quorum Books, for a speedy review of the manuscript and a prompt editorial decision concerning its publication. Terri M. Jennings, from Jennings Publishing Services, provided superb editing.

As with any work that stresses applications, the debt of the author to previous researchers in the fields included in this monograph is substantial; the mistakes, however, are only mine.

Finally, I thank my family for their understanding when the development of some parts of this book, especially Chapter 5, stole much of our time together.

This book is dedicated to those who help me.

# MODELING

This chapter provides an overview of the modeling process in general and simultaneous equations systems in particular. The intention is to set the framework for the study of approximate equations within the context of problems selected for this monograph.

The presentation of modeling will be rather general, if not interdisciplinary, in order to be at least informed of developments in areas that may influence research in approximate equations systems. The chapter will review the concept of equations first, and then study modeling in general. Finally, it will focus on simultaneous equations systems. The analysis of simultaneous equations systems will reveal their limitations, some of which are addressed by approximate equations systems (AES), as will be seen in Chapter 2.

The presentation will start from scratch but may become quite detailed in one or more areas. Readers familiar with calculus and simultaneous equations should have no difficulty following the discussion. For those who are less familiar with those mathematical objects, some texts in either one of the areas of linear algebra, optimization, or mathematical methods in economics are recommended. Suggestions are given at the end of the chapter.

## 1. EQUATIONS

The term "equation" brings to mind the verb "to equate"; both of which make reference to establish a relationship of equality between two or more things. Equating is an instance of comparing, which is one of the cognitive skills we use most often. We could say that making decisions rests largely in comparing alternatives.

In mathematics, an equation is a relationship between two variables, as in $2x = z$, or between a variable and itself, as in $x^2 - 3x + 2 = 0$. The latter is an equation of x in the second degree because that is the highest power to which x is raised. The former is especially useful when comparing two variables, the latter when studying the behavior of a single variable. If we know the value of one variable in $2x = z$, we can easily calculate the other. The algebraic form of $2x = z$ is the symbolic expression $[ax = bz]$, where a and b take the place of actual values. Algebraic expressions allow easy and general manipulation of mathematical terms.

We build equations for learning, and causality is a cornerstone of the edifice of learning. The equation $2x = z$ acquires new and valuable meaning when we can posit that z causes x. The workings of the variable z may be unknown to us, but if we can measure it or expect it to take some specific value, then we can calculate x. For example, if $z = -4$, then $x = -4/2 = -2$. We try to add further detail by relating a family of caused variables $(x_1, x_2, x_3, \ldots, x_k)$ to a family of causing variables $(z_1, z_2, z_3, \ldots, z_k)$, both families having the same size "k"— that is, members. However, every time we add a new pair $\{x, z\}$ of variables, we will need a new equation. Each equation is a piece, or strand, of information we weave into others to structure knowledge. Systems of equations with two or more variables are called simultaneous because the variables find their solution value at the same time.

A simultaneous equation system, $[Ax = b]$, is a complete collection of relationships among variables that fully describe a phenomenon or event. These relationships are precise and the variables precisely measured. Further, we are able to separate variables that are determined within the system $(x_i$'s) from those that are independent $(b_i$'s). Once the variables are determined, a simultaneous equation system can be expressed in different ways. For example, a two-dimensional (two-variable, two-functional relations) system can be expressed in matrix form as:

$$(1) \qquad \begin{vmatrix} 1 & -2 \\ 1 & 2 \end{vmatrix} \quad \begin{vmatrix} x1 \\ x2 \end{vmatrix} = \begin{vmatrix} -4 \\ 8 \end{vmatrix}$$

$$1\,x_1 - 2\,x_2 = b_1 = -4$$
$$1\,x_1 + 2\,x_2 = b_2 = 8$$

or, as written earlier,

$$(2) \qquad Ax = b$$

where the two-by-two matrix of coefficients is called A, the vector (2 rows by one column, $2x_1$) of dependent variables is expressed by lowercase x, and the vector of independent (or causing) variables is expressed by lowercase b. It is customary to represent matrices and vectors by capital and lowercase letters, respectively, and constants and variables by the first and last letters of the alphabet, respectively. Therefore, we can understand that A represents a set of constants or parameters because we know them well; b represents variables that happen to have a given set of values, but we may not know as well. Furthermore, some or all of our variables may have probabilistic meaning, which makes it necessary to amend the expression A x = b accordingly, as we shall note later. This book will concentrate on systems such as those in the equations (1)-(2), where all the variables and parameters are real valued numbers in the hyperspace, or multidimensional space, of real numbers $\mathfrak{R}^k$.

Simultaneous equation systems are usually expressed as A x = b, which will be the chosen expression throughout this monograph.

Simultaneous equations are a generalization of the most common and natural arrangement of things: a two-dimensional arrangement of a single class of objects. For example, consider sales within a firm. The firm may sell different products—or a product of different types—in different markets. The two, logically coherent sets, are $x_i$'s and $b_i$'s and form a class—the classification itself—of our objects. A matrix appears naturally when we arrange, classify, or organize the information about sales by relating the set of products = x = {$x_1$, $x_2$} to the set of markets = b = {$b_1$, $b_2$}, through the structure A, as in $x^* = A^{-1} b$, with which we can build a table of relations

|       | $b_1$  | $b_2$  |
|-------|--------|--------|
| $x_1$ | 0.5    | 0.5    |
| $x_2$ | -0.25  | 0.25   |

where we can determine the contribution, impact, or dependence of each of the xi's on the variables bi's. For example, the table above says that

$$(3) \qquad \begin{aligned} x_1 &= 0.5\, b_1 + 0.5\ b_2 \\ x_2 &= -0.25\, b_1 + 0.25\, b_2 \end{aligned}$$

which means x1, sales of product x1, which totaled 3, were generated 50% from each market. Negative values can have the interpretation of returned merchandise. The coefficients above could also take the interpretation of prices; or "yields" coefficients in an input-output (b $\rightarrow$ x) production model, with the cross-coefficients ($a_{12}$, $a_{21}$) representing technical substitutability relationships; or allocations (expenses, $x_i$'s) in a two-income ($b_1$, $b_2$) household.

From equations, we can also construct in-equations, which are relations of inequality. These relationships play a major role in mathematical programming and other areas of optimization theory, see Beckenback and Bellman (1961) and Nemhauser (1988). Equations are the most pervasive tool we use, implicitly or explicitly, with the exception of natural language. Even with natural language, we reason by forming relationships. It is important to keep in mind throughout this monograph that equations are information. This process of gathering, arranging, and organizing information (in either numerical or linguistic form) is how we come to understand the world.

At this point, we could move directly into simultaneous equations systems and approximate equations systems, but that would be like discussing the trajectory of a plane by examining a single picture. We need to know the origin to assess its direction and, if necessary, modify its course. For a similar reason, we need to review the motivation and foundations of the information management process we call modeling before proceeding with the discussion of SES and AES.

## 2. ON HOW AND WHY WE MODEL

We reflect and think about what comes to us and how we may avoid unwelcome events. We do this because we are intelligent creatures. The ability to imagine alternative situations, represent them via symbolic or natural language, and reason about things—intelligence—is part of our endowment of resources and thus it is also part of being human.

When our thinking aims at something specific, such as events or problems that concern us, we find ourselves using structures of thought we could call "forms of intelligibility." Simultaneous equations systems, mathematical programming, an Income-Saving/Loans-Money (IS-LM) model, predicate calculus, and so on, are all forms of intelligibility. But so are prose, poetry, painting, dance, the concept of "eternal return," the Tibetan "Book of the Dead," and initiation rituals. They all try to understand what happens around us and how we can improve upon our predicament. They all try, in other words, to "ward off the terrible chaos"—Deleuze and Guattari (1994)—of our ignorance and (perhaps) liberate us from tormenting uncertainty.

Many of our intelligibility forms mutate from language into numerical mathematics, as the larva transforms into the chrysalis, and it is not clear where the transition point lies. Moreover, forms of intelligibility are also forms of communication. Over time, however, we have developed methods of knowing that we call scientific. These aim at clarifying both what is known and unknown. They stress (1) communicability, that several observers will observe the same under the same conditions; (2) objectivity, in the sense of the findings being independent of any researchers; and (3) replicability, if two researchers claim to have found cold fusion, then another two, or the same two should be able to replicate the event.

We can trace fragments of the scientific method to Greek philosophers, who hypothesized that learning is linked to happiness (Socrates), that existence reaches a point of repose (Parmenides), or one that is always in flux (Heraclitus), that there is a "form," ideal, representative of things in and out of our head (Plato); and that things follow a path to completion or perfection (Aristotle). Aristotle also developed logic in his search for a general-purpose tool to acquire knowledge. He employed terms such as quality and quantity, induction and deduction, cause and effect, and investigated the formation of perfectly defined concepts or "categories."

Descartes (1596–1650), Locke (1632–1704), Hume (1711–1776), and Kant (1724–1804) contributed greatly to developing the method of knowing we call scientific. What was at stake in Descartes' time was the formalization of a system by which men could gain knowledge through means other than revelation. Locke explained how we come up with ideas without the help of a transcendental entity, by observing the external sensible objects and the internal operations of our mind. Hume noted how we can make thinking mistakes and how many of these—meaning most of them—can be avoided with a healthy dose of skepticism. Kant's program is considered fundamental in Western thought; he furthered the work of Plato and Aristotle on logical analysis. Kant noted how our mind tries to make sense of all the information we receive and everything we experience. He showed us the complementary aspects of rationalism and empiricism.

This activity is carried out by "a priori" forms of sensibility (space and time) and also "a priori" categories of judgment, such as quantity (unity, plurality, universality), quality (affirmation, negation, limitation), relation (substantiality, causation, reciprocity), and modality (possibility, actuality, necessity). These a priori items are not concepts, nor do we derive them from experience. They exist in our minds as part of our resources to understand the world around us. It is not difficult to see many of these a priori forms at play in the building of a simultaneous equation system. Our current methodology of science owes much to Kant, who also made contributions on ethics, aesthetics, and metaphysics. The process of developing a simultaneous equations system, such as $A x = b$, fits Kant's framework: we use experience and observations to record data and we reason out a model.

Descartes (see, for example, 1968) was after a dynamic method of learning that could be applied to itself (recursion) to improve itself: an epistemology. His four famous rules were (1) intuition, (2) analysis, (3) synthesis, and (4) deductive reasoning, which links first principles and ultimate consequences. Each of these principles has direct import on the development of simultaneous equations systems. We first understand that a problem (financial planning, the macroeconomy, or portfolio selection) can be put into a system such as $A x = b$. Then we analyze exactly why we can do so (e.g., as John Maynard Keynes did in his *General Theory of Employment, Interest, and Money*), by justifying and developing a system of relationships among variables. By synthesis, Descartes meant the clarification and organization of the simple into the complex. For example, $A x = b$ may be part of a more complex $Z b = h$ system. We could arrive at the more complex system only after learning about the simpler one. Descartes's fourth rule tried to emphasize how reasoning and memory can be used to link first principles (h, above) with final consequences (x). Note that, in our example, we could make that link explicit by performing a straightforward operation: $A x = y = Z^{-1} h$. This implies $x = [A^{-1} Z^{-1}] h$.

A method to learn needs objects. As we were learning about epistemology, mathematical objects experienced a most spectacular growth. The art of counting (arithmetic), the use of symbols for numbers (algebra), and measuring and representing figures (geometry) made possible the development of calculus and analysis (the study of the objects utilized in calculus) by Gottfried Leibniz (1646–1716), whose notation we still use today, and by Sir Isaac Newton (1642–1727), who was able to provide a comprehensive and acceptable explanation of mechanics in our world. Fermat (1601-1065) and Blaise Pascal (1623–1662) developed the mathematical theory of probability. Fermat and Descartes wedded algebra and geometry—see Kramer (1981), p. 137; and it was the French mathematician François Viète (1540–1603, Vieta in Latin) who advanced our understanding of simultaneous equations and their solutions.

The two major streams in mathematics, algebra, and calculus, converged during the time of the aforementioned researchers. Our knowledge of simultaneous equations systems later progressed with contributions of Agustin Louis Cauchy (1789–1857), who set the logical foundations of calculus, and with the research on calculus and the theory of real numbers by Nikolay Ivanovich Lobachevsky (1793–1856), Peter G. Lejeune-Dirichlet (1805–1859), who provided the basis for the modern use of functions, and Julius W. R. Dedekind (1831–1916). Among Carl Fiedrich Gauss's (1777–1855) many contributions to mathematics, there are procedures to solve simultaneous equations such as a simple method (Gaussian elimination), which can be modified to provide the inverse of the matrix A (Gauss-Jordan elimination), and an iterative procedure (Gauss-Seidel method). Gauss also developed the method of "least squares" to fit a series of observations on a variable x to those of a variable y by solving a SES system known as the "normal equations."

As we shall see later in this chapter, a system of simultaneous equations such as $A x = b$ also appears in the optimization of a quadratic form. This unites the areas of linear algebra with those of calculus. Joseph-Louis Lagrange (1736–1813) developed some of the results we will use in our study. For example, a method to solve a quadratic (second-degree, nonlinear) form subject to a linear restriction (or a quadratic restriction subject to a linear function) by what are called "Lagrange multipliers." This method is widely used and preferable to quadratic programming because it requires fewer computations. We will use Lagrange's method in our portfolio optimization example in Chapter 5.

The effort in developing mathematical objects mirrored the effort in learning about reasoning and epistemology, which are areas that blend in the work of Russell, Frege, Carnap, and others. Russell noted that what matters most in problem solving is the quality of our reasoning. The choice of particular objects (language, symbols, numbers) should be secondary. Of course, some objects are preferable to others in certain cases, but still the quality of thought is what counts. For example, when planning to farm a field it is best to use quantitative relationships among fertilizers, seeds, and watering cycles. However, when a family psychologist tries to address a dysfunction, qualitative methods may prove more useful. In sum, mathematics and logic, or any other area of learning, are about thinking.

The reader interested in the development of mathematics is referred to the excellent text on the growth of modern mathematics by Kramer (1981), and the useful accounts on mathematics and its uses by Kline (1967) and Courant, Robbins, and Stewart (1996). For an entertaining and unconventional account of the history of calculus, the reader is referred to Berlinski (1995). We now turn to the description of models and the modeling process.

## A. A Model and the Modeling Process

A model is a more or less simplified representation of a fragment of reality. It consists of a collection of mathematical objects with their associated syntax (rules of operation), a set of theories that provide the semantic (meaning), and data that flow through these objects to provide the information we seek. Of course, the objects may be those of natural language, which sets off a wide range of possibilities as well as challenges.

In the case of a simultaneous equations system such as $A\ x = b$, the objects are those of linear algebra and calculus, that is, matrices, real valued variables, and relationships. Theories justify using those objects; for example, a variable representing private investment is thought to be negatively related to another representing interest rates. We always assume we will find the data we need; however, we must often build models and theories based on what is available.

The process of developing a model can be described very simply. We first notice a problem and try to determine the main factors involved. Then, we try to specify variables that measure each of the different factors at play in our event. We further separate these variables into causing (exogenous, determining, independent, external, etc.) and caused (dependent, endogenous, determined, internal, etc.). This brief account reveals that modeling, in general, depends on the following three conditions:

1. **Identifiability** or distinguishability. We must be able to identify the problem to begin with, which may not be easy to do.
2. **Measurability**. This is a "sine qua non" requirement in quantitative modeling. It is often said in business that one cannot manage what cannot be measured and that the first step to managing a firm is to start counting things (clients, the money they spend, the time they stay in the store, the time it takes to make a sale, square feet of inventory, etc.). After a while of doing so, good ideas will come. Measurability is particularly crucial in a simultaneous equations system model.
3. **Representability**. What we do has to be related to the problem that we are trying to solve. This observation seems sensible to the point of being redundant. However, it is heartbreaking the time and effort we place in models that have no relationship—at all—to reality. The worst of all is that we know it, and do nothing about it. Economists owe much of their bad name to their insistence on using "general equilibrium" models in a most discontinuous, disequilibrium-prone world, or to applying "rational expectations" models to countries without monetary systems or sophisticated labor markets. (By the way, both of these models make use of simultaneous equations systems.)

Simultaneous equations models such as $A\ x = b$ are rather impressive in both their potential and in their requirements. Moreover, many other modeling options, such as linear programming, build upon the structure of simultaneous equations systems, as we will see in the next section.

### B. Considerations on Modeling

In spite of centuries of hard work at learning, Alfred North Whitehead once noted that Western philosophy is simply a footnote to Plato. This is a very profound comment, which took us years to fully understand. What Whitehead meant was that all we have are representations of reality, in most cases abstract "ideals." Berkeley may have added that those ideals exist only in our minds, and that they are not "the real thing."

Developing satisfactory models that explain and predict real events is still very difficult, in spite of developments in analytical techniques and computation. First, we face the overwhelming complexity of reality itself. It seems the more we learn, the more we need to know. Second, not everything is quantifiable. Even for some of the quantifiable events (e.g., short sales of stocks) theories and solution procedures may be lacking. Third, to generate words, concepts, and models for problems is not the same as solving them. For example, there is considerable research into ascertaining the role played by language in modeling. We must recognize, however, that language is all we have to form our representations in some problems. A fourth element that makes model building challenging is its human component.

Taking into account the first three issues above, our modeling of each problem or phenomena could be explained according to a matrix, or framework, relating the phases of problem solving (knowledge representation, solution procedures, cognitive maps), to the elements of a model (objects, theories, and data). For example,

|  | Objects | Theories | Data |
|---|---|---|---|
| Knowledge Representation |  |  |  |
| Solution Procedures |  |  |  |
| Cognitive Maps |  |  |  |

We have advocated for this framework earlier, Tarrazo (1997e), because it allows the study of practically every problem appearing in textbooks, whether they are quantitative or qualitative. This framework was inspired by readings on cognitive science, where the "representation" nature of every model, and the subject-object relationship are stressed. Note that the problem of assessing the future is studied under the heading "cognitive maps," which permits a wider interpretation and treatment than that allowed by the conventional "prediction" problem. Exhibit 1.1 presents an outline of the modeling process.

The fact that our models deal with humanistic systems brings at least two additional considerations: the subject-object relationship (a Kantian theme), and the general issue of accuracy. Exhibit 1.2 presents two interpretations of the subject-object relationship. Picture A depicts the subject, the investor, as an island in the investment market. Picture B highlights the role of the model, or representation, as a mediator between the subject and the object. This picture is consistent with what we mentioned in the first paragraph of this chapter:

individuals build models to modify their environment. In other words, the model is both a picture of the world and our tool to change it.

**Exhibit 1.1**
**The Modeling Process**

1. Need to do something
2. Notion of determinism
3. Concept (classes, categories, and so on) building
4. Structuring of concepts:
   A.    Conceptual hierarchy
   B.    Causation channels
5. Selection of the modeling domain
   A.    Knowledge representation
   B.    Solution procedures via operations allowed in such domain.
   C.    Cognitive maps (prediction)
6. Decision-making
7. Learning

*Source*: Tarrazo (1997g).

When we deal with complex problems in humanistic systems the problem of lack of precision seems inevitable in all but the most trivial formulations. This situation has been analyzed by Zadeh, the "father" of fuzzy sets, who formulated his famous "incompatibility principle": "As the complexity of the system increases, our ability to make precise and yet significant statements about its behavior diminishes until a threshold is reached beyond which precision and significance (or relevance) become almost mutually exclusive characteristics" (Zadeh, 1973).

Exhibit 1.3 provides a visual presentation of the trade-off between precision and generality for any model. When we generalize a model to cover many situations or serve several subjects, it must become more general and abstract, which brings imprecision. In the same manner, when we try to extend the time frame for a model, some of the precision that was used to fit each temporal idiosyncrasy or behavior is masked by its long-term patterns. For example, a model that represented interest rate volatility dynamics in the United States fairly well before 1980 and after 1983 would have trouble dealing with high interest rate volatility of the 1980–1983 period. The model that may represent well the stochastics of the 1980–83 period may overrepresent dynamics outside that period.

And there is the human factor, about which Caws notes the following self-explanatory comment: "The outcome of mediation by a mind is likely to be a complex object unlike any other object, unless constrained by simplifying externalities or deliberately standardized" Caws (1997, p. xxv).

**Exhibit 1.2**
**Representation of the Investment Decision**

Picture A

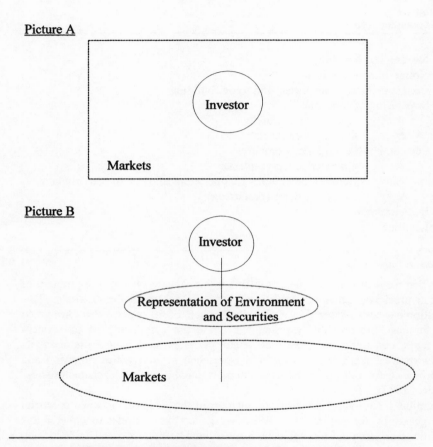

Picture B

*Source*: Tarrazo (1999). © Institutional Investor Journals. Used with permission.

### C. Other Approaches to Model Building

Our review of modeling as a representation of phenomena has not examined all the current modeling practices or approaches, but only those with a direct relationship to the main focus of the monograph, that is, those related to simultaneous equations. A brief overview, however, of other modeling choices will be useful to close our general presentation of modeling.

Calculus and probability methods provide exact and stochastic simultaneous equations methods. Calculus methods also provide mathematical programming models, some of which can also be derived from linear algebra. However, these are not the only ways to model. As Exhibit 1.4 shows, the first decision when choosing models refers to using symbolic or natural language. Within symbolic language, we can choose the tools of the algebra of sets (classes, relationship operations, and so on), which are well suited for qualitative analysis, or those of calculus and probability, which are better suited for quantitative analysis. Within quantitative analysis, we can choose between causal models, that is, models that explicitly state causal relationships (for example, simultaneous equations models) or those that do not. The latter are called connectionist models. Neural networks are an example of such models, and they are starting to seriously compete with conventional models because of their ability to mimic data patterns.

**Exhibit 1.3**
**Information and Precision**

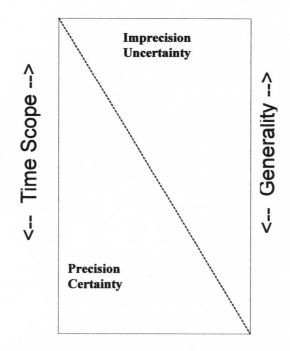

*Source*: Tarrazo (1997g).

Formal modeling with natural language has intensified since the times of "analytical" philosophy. At first, much of our knowledge took the form of texts, and researchers will carry their analyses mostly in discursive form. Propositional calculus grew out of the need to clarify what was being said, but it seems to impair the message as well as the medium. Artificial intelligence represents one of the latest efforts to use natural language in our modeling endeavors.

We will revisit this Exhibit 1.4 in Chapter 6, where we will also compare some of these modeling options to approximate equations systems.

**Exhibit 1.4**
**Methodology Shopping List**

| Information | Methodology |
|---|---|

*Symbolic*:

| Qualitative | Algebra (Boolean, lattice, fuzzy sets-based, etc.), it is always causal. |
|---|---|

| Quantitative | Causal: |
|---|---|
| | Calculus |
| | Mathematical programming |
| | Probabilistic |
| | Noncausal, or connectionist, modeling: Neural networks (some neural nets are causal). |

*Linguistic*:

Discourse analysis (inductive)
Propositional calculus (deductive)
Artificial intelligence
    Symbolic (schemas-based)
        (inductive)
    Procedural or rule based    (deductive)

---

## 3. SIMULTANEOUS EQUATIONS SYSTEMS

This section is dedicated to simultaneous equations systems. It could have been entitled "A dream come true." Imagine that you face a difficult problem that, nonetheless, can be perfectly specified. Imagine that one can find a number of variables that fully represent the problem, for which data is available to a limitless degree of precision, and that we can further split into dependent and independent variables. Imagine that we can determine the precise way in which the variables interact via functional relationships, and that the resulting system can be solved nontrivially; that is, we can find a solution value for each of the dependent variables that is compatible with the values of the exogenous variables and the functional relationships in our system. This scenario would represent a "dream" for any scientist in the social sciences, yet this is the scenario described by a simultaneous equations system such as the one we introduced in the very beginning of this chapter.

Our dreams in this case have been fueled by spectacular successes in other sciences such as physics. Representations of the real world are possible with objects that resemble SES, but that are more specialized, such as, for example, the tensors of Albert Einstein's theory of relativity, and the type of stochastic matrices employed by Richard Feynman and others in quantum electrodynamics. In economics and finance, however, perhaps because we deal with humanistic systems, things work differently. We are still trying to make the best out of the generic SES.

## A. Importance in Economics and Finance

Why are SES important in economics and finance?

Much of the economics research at the macro (aggregate) level tries to improve our ability to influence the economy, especially by government actions such as monetary and fiscal policy. It would be ideal to be able to represent the economy using a SES system. The variables of national income, exchange and interest rates, and employment as dependent variables, and government policy variables (expenses, tax revenues, and money supply) as independent variables. Then, the planner would adjust the policy variables to guide the economy through "thick and thin." Of course, as we shall see in Chapter 3, the reality is that we are not even sure if government variables can be taken as exogenous, or if agents use the static representation implied by traditional SES models.

In finance and the microeconomics of financial markets, SES systems can be used to obtain equilibrium prices as a relationship between supply and demand, or equilibrium portfolios in a contingent states pay-off matrix, see Henderson and Quandt (1980), Varian (1984), and Huang and Litzenberger (1988).

SES structures are also present in finance and accounting in a very special way. In accounting, many critical relationships such as those of budgets and balance sheets depend on identities between sets of variable inflows/outflows of budgets, assets, and liabilities in balance statements. Other financial statements, such as the statement of cash flows and the change in stockholders' equity also reflect identity relationships between sets of variables. Accounting and finance are related in many ways, but it is instructive at this point to observe the following:

|  | Accounting | Finance |
|---|---|---|
| Elements | categories | variables |
| Objects | identities | equation system (simultaneous or not) |
| Reference | past | future |
| Field | legal | economic |

The elements of accounting are well-defined concepts (categories), that must observe relationships of identities if the books are well constructed. They refer to past activity and, thus, can function properly as legal records. The elements of finance are variables, often borrowed from accounting at the risk of compromising their required forward-looking nature. The reference area is the future and the value should be mostly economic rather than legal. The preparation of pro-forma statements represents the bridging activity between the two business disciplines.

In microeconomics and those parts of finance related to individual choice, SES offer the possibility of representing choice, because in the usual $A x = b$ specification, A can be taken to represent the coefficient of response of the individual, b the surrounding environment (Picture A in Exhibit 1.2). In sum, a SES model clearly shows the key element of choice: the difference between what can be changed and what is given to the individual. The contextual interpretation of an SES model is summarized in Exhibit 1.5.

In a way, we could say that economics and finance have been immersed in what we would call "The Age of Calculus," as evidenced by the centrality of SES models in textbooks such as Chiang (1984), and Samuelson (1983). One may think we can overcome SES models by replacing them with more sophisticated constructs; but when we study this option—see, for example, Duffie (1992, 1998)—the signpost signaling to the road "more complicated" seems to lead us to a place called "nowhere."

### B. SES and other Mathematical Procedures

Many other models build upon the structure of a simultaneous equations system. It is important that we realize this because it shows the potential of our study on approximate equations. A system such as $A x = b$ normally provides a unique set of values for the dependent variables if the matrix of coefficients, A, has an inverse—its highest order determinant is not equal to zero, and the order of this determinant (rank of A) is k. In some cases, we understand those values to have special meaning (equilibrium, optimal values, and so on). What they indicate are those values of $x_i$'s consistent or compatible with the values of A and b, that is, the set of relationships $A x = b$ can be interpreted as a (full) set of restrictions. In this context, existence of an inverse indicates a restriction per dependent variable, and all the restrictions are binding or effective. It is easy to see that if we eliminate some of the equations in $A x = b$, the system may admit multiple solutions because it would have more unknowns than equations. This is exactly the problem tackled by linear programming.

The basic linear programming problem can be written as

(4)          Max          $p = c x$
             subject to    $A x = b$

where p is a scalar, c is a (1xk) vector, A has a rank lower than k, and x and b are both (kx1) vectors. The function $p = c x$ is called the objective function and signals which of the $x_i$'s are most important—that is, ranks them. If A had full rank, we would not need the objective function at all. The simplex algorithm is the most efficient way to solve these types of problems.

A simultaneous equations system is also generated in the optimization of second degree nonlinear functions, since their first derivative provides a linear form such as A x = b, see for example Chiang (1984), Luenberger (1984, 1998), Foulds (1981), or Boot (1964).

Programming offers several potential advantages over SES such as the use of non-negativity restrictions and the ability to find solutions when the matrix of restrictions is not of full rank (limited information). It is also very helpful when variables are restricted to only integer values—dealing with integer SES (Diophantine analysis) quickly becomes very difficult. Programming, however, also presents some limitations when compared to SES. It applies to mostly static problems, since dynamic programming is cumbersome, and one can only study changes in one exogenous variable. Postoptimal analysis in SES and programming systems is comparable in terms of difficulty and yields in SES and programming models.

The programming problem adds to the SES treatment of choice and policy problems. In the programming case, the planner faces fewer restrictions than in SES. This creates some flexibility in decision-making but also highlights the analysis of contingencies. See Exhibit 1.5.

Current research in optimization techniques is bringing programming and SES specifications closer in their foundations, solution procedures, and in their goals, which seek to (1) add flexibility to conventional specifications, (2) expand the scope of traditional models, and (3) lay out a bridge between SES and programming methods and novel optimization techniques such as lattice, graph, network, inequality, and polyhedral theory. Reviewing the excellent texts by Lai and Hwang (1993), Luenberger (1984), and Nemhauser and Wolsey (1988), provides a fresh and revealing overview of the current state of optimization theory and practice.

Dynamic considerations can be added to our customary SES model. First, because a SES may be understood as an instance of a more complex—for example, dynamic—system. This is what Samuelson (1983) called the "Correspondence Principle." Second, because one can use the calculus of variations to study problems of dynamic control in closed systems, see Pierre (1986). Finally, multivariate specification of dynamic systems (difference and differential equations) makes use of the SES models, see Barnett (1992) and Drazin (1992).

A distinction is often made between (static) deterministic systems such as A x = b, versus (static, and ergodic) systems such as A x + $\xi$ = b, where $\xi$ represents a vector of probabilistic errors. In Chapter 2 we will compare these objects with those of approximate equations systems. In general, however, we will concentrate on nonprobabilistic systems because we do not want the veil of probability to hide our ignorance.

Calculus and optimization theory offer a wealth of procedures to solve problems but, not surprisingly, in economics and finance, in addition to sophisticated techniques and objects, we always seem to need something else. For example, a treatment of the investment problem such as that by Luenberger (1998), flawless as it is in terms of optimization, still leaves the reader with the impression of missing critical parts of what investments are all about. The same can be said of financial planning, macroeconomics, and portfolio management. Once again, the signpost signaling toward the road "more complicated" seems to lead us "nowhere."

**Exhibit 1.5**
**Contextual Interpretations of SES and Mathematical Programming**

|   | Calculus/Programming | Economics | Planner |
|---|---|---|---|
| A | Structural relationships | Parameters | Response coefficients |
| x | Dependent variables | Policy variables | Decision variables |
| b | Independent variables | Exogenous variables | Environment |
| p | Objective function | Signaling when A does not have a full rank | |

Our intuition is that progress in economics and finance may come from approximate, perhaps customizable, SES-like structures rather than from sophisticated general-purpose, comprehensive models.

### C. SES as a Structure of Intelligibility

SES models are our most sophisticated structure for intelligibility. That is why they are so important. We have already said that a SES is a form of intelligibility, but the importance of the concept of "structure" is very rarely well understood.

Structure (from Latin) has a similar connotation to that of system (from Greek) and is a self-contained set of relationships among elements of logically related sets. Structuralism, says Caws, "(I)s a philosophical view according to which the reality of the objects in the human or social sciences is relational rather than substantial." (1997, p. 1). In other words, in social sciences such as economics and finance, establishing a representative or self-sufficient subset of relationships may be more important than specifying each of its elements precisely. For example, one may have a blurry copy of a portrait, yet still identify the subject by focusing on relationships. This is, actually, all we do in economics and finance. We try to identify "faces" of the economy, of markets, of our future finances based on blurry pictures.

Economic (and noneconomic) policy makers always have a blurry picture of their country's economy and still manage the country. Executives have blurry pictures of their firms and their markets, and yet they continue to manage their firms. The best example is that of the portfolio manager, who may reach her objectives in terms of risk-adjusted returns by focusing on the group of securities, rather than on specific companies. Portfolio theory is simply an implementation of structuralism: focus on the group, so individual elements become less critical, perhaps even expendable!

In other fields of learning, such as biology, or psychology, which may be taken to be "harder" or "softer" than economics and finance, the mathematical weight of a matrix system may be diminished, but the concept of structure and relationship is not. Providing even a few examples of how SES are used in psychology would distract us from our main train of thought at this point in the study. The interested reader is referred to Bolles (1993), and Ansbacher and Ansbacher (1964), where the following expressions can be found, among others: Stern's matrix of goals, Maslows' matrix of needs, Lewin's motivational vector, the structure of stimuli-responses (S-R), and so on. Moreover, these structures are not necessarily static. For example, Tolman criticized the static S-R model because it did not include expectations. This sounds very familiar to economists, the S-R model could be modeled as $A r = S$, in which we make the response a function of the stimuli, given the behavioral matrix A. Expectations could be included by "dynamizing" $A r = S$. The emphasis of the "Gestalt School" on structures is obvious.

The reason to focus on "blurry pictures" is that we have no other choice; we can only come up with representations of phenomena. But, again, it is the structure that saves the day because, as Caws (1997) notes:

1. "Each scientific statement can in principle be so transformed that it is nothing but a structure statement." (Caws, 1997, p. 176.)
2. "The amassing of evidence never yields the hypothesis, which always comes from a structural insight that organizes the evidence that has been amassed." (Caws, 1997, p. 128.)
3. "If structure is a system of constraints we must look for the origin of the constraints elsewhere than in the structure; but this need not deprive the structure of its explanatory power, because it explains synchronic features rather than origins." (Caws, 1997, p. 52.)

We must aim, then, for uncovering structures because they are the fabric of our knowledge, and because "(I)t is the structuring activity of the mind, its appetite for interrelations among things, that lends intelligibility to the world, or of things in the world, we think of ourselves as inhabiting or having dealings with. For it is possible to think of the intelligible not as given, but as constructed" (Caws, 1997, p. 37 and p. 159.)

In sum, dealing with single items, as we often do, is very difficult. We know we cannot deal with the whole because it is hard to identify and capture. Our best approach, therefore, and often the only possible one is to handle the smallest set of items that form a relational complex—Ockham's razor. In general, we are tempted to say that the best we may have in modeling is the crafting of plausible, even if incomplete, representational structures.

### D. Problems with SES in Economics and Finance

We have not yet taken advantage of the opportunities offered by SES; but why not? This is because we have been using SES in the wrong way. We have focused on SES strictly as mathematical objects. We have sought to adapt our reality to their needs instead of adapting them to the realities of social sciences. We have not realized the implications of Zadeh's principle, nor have we emphasized what is most promising in a SES model: its ability to model causal relationships.

The calculus world of SES requires a precision we do not have in social sciences. However, there is a bridge from the exact wonderland of calculus into our "terra incognita" that preserves for our models the scientific characteristics of measurability, distinguishability, and representability. We build such a bridge by viewing SES systems as approximate.

This bridge also reaches into other areas such as qualitative analysis, but that is another story. We can say, though, that approximate simultaneous equations systems are also a way to take advantage of information that is mostly of a qualitative nature and would be wasted if the model were to take an exact form. By relaxing the conceptual requirements of conventional mathematical objects, we bridge the gap between major divides in scientific methodology.

By recognizing error, the hard, accurate, observable mathematical variables become flexible and real. The exact, precise relationship becomes more flexible and realistic. In other words, the model becomes practical and becomes customizable simply by including errors.

Moreover, we have not used SES models in economics and finance optimally because we have forgotten that models are the representations that individuals use to make decisions, as was depicted in Exhibit 1.2. Individuals need their own models as much as they need custom-made suits—no need to rule out "prêt-à-porter" models yet—and this could be accomplished with AES. If we eliminate some equations or variables in a SES model, we may irreparably harm the model. However, taking into account the "error-carrying capability" of the overall system, we might develop a simpler, custom specific AES model.

In the next chapter, we will focus on AES and uncover further motivation to use them in practical endeavors. In the last part of this monograph we compare AES to other methods of inquiry.

## 4. MODELING, SES AND AES: PRELIMINARY ASSESSMENT

In the near future exact specifications will be replaced by interval ones, especially in the social sciences.

It is necessary to note the general limitations of conventional or standard modeling in order to appreciate the role AES may play in the near future. In the next chapter, once we have studied AES themselves, we will have the opportunity to assess their potential more fully.

We could organize the general limitations of our current modeling practices into the following three headings:

1. Limitations of a first kind. These are those limitations inherent in the objects, those of the theories, and those caused by absent or inadequate data. For example, students review microeconomics in business management courses. However, microeconomics focuses on data that are hard to obtain from accounting, about firms that are in perfect competition, on perfectly divisible inputs to productions, and on financial markets simplified to the point of having a single interest rate, which cripples the usual and critical short term-long term financing decision.
2. Limitations of a second kind. These are those caused by our methodology choices. For example, when we choose to model something with calculus instead of qualitatively; when we misperceive or misapply causality; when we confuse the object-subject relationship; or when we confuse linguistic meaning with mathematical value. In sum, some of the limitations of the second kind stem from cognitive considerations, others from the nature of the model as a representation as Foucault (1994) and Rorty (1979) have noted, among others.
3. Limitations of a third kind are those of a metamathematical and ontological nature. That is, those related to intelligibility proper; whether we can actually replicate reality using logical tools—Godel's impossibility theorem—and whether that is even relevant.

Some of these limitations will haunt our modeling efforts for quite some time. Fortunately, limitations are also opportunities to overcome adversity, and opportunities to try new approaches and new objects such as AES.

Because of what we have presented in this chapter, we can say that AES can address some of the limitations of the first kind; for example, build a bridge between exactness and inexactness, provide space for theoretical errors, incomplete models, and also provide a shock absorber for errors in data collection and transmission.

Modeling with AES can also address some of the modeling limitations of the second kind; that is, those related to misperceived causality, blurred borders between subject-object and determined-determining factors, and so on. With AES, parameters and functional relationships become realistic because they give up their pretension of accuracy and thus transform into structures of intelligibility. Given this, we may opt for building local representations instead of global ones.

AES cannot address limitations of the third kind. They do, however, allow us to take formal models with a grain of salt. The models we build are, at best, *only apparently valid representations of perceived facts*.

Simply adding errors to a set of simultaneous equations systems will not do much for improving our knowledge of the world. The real prize is using approximate equations systems to improve upon our structures of (subjective) intelligibility, to free ourselves from the limitations of our tools by taking into account their "error-carrying capability" and filling that void in the model with our intuition. This is what makes us unique as decision-makers. We want the reliable car, and some room for maneuvering.

In the next chapter, once we have presented AES properly, we will understand why there are numerical, probabilistic, theoretical, informational, behavioral, and cognitive reasons to use AES.

## SUMMARY

In this chapter we have examined modeling in general, its principles, and noted some of the limitations of conventional modeling practices.

We have seen that SES are at the core of current quantitative modeling and upon which other methods such as linear and nonlinear programming are based. The same characteristics that make simultaneous equations systems central are also their limitations. SES are valued because of their comprehensiveness and completeness in terms of well defined and well developed mathematical objects and their ability to include theories, data, and provide testable hypotheses. It is fair to say we have not yet taken advantage of the opportunities offered by SES. What may be preventing us is simply the insistence on levels of accuracy and precision that we may not have.

The calculus objects we employ in SES require more accuracy than we have in practice. In a more general sense, it should be obvious that what we know in each problem is less than what we ignore. We would not have a problem if it were otherwise. AES offer ways to take advantage of the strengths of SES modeling while factoring in some critical modeling limitations and uncertainties.

## FURTHER READING

Our manual in approximate equations can be used as a sounding line to determine the depth of the reader's mathematical knowledge. Moreover, this monograph is more of a starting point in the reader's incursion in approximate methods. We recommend reading this book first and, if absolutely necessary, those with less mathematical background could consult the references below.

The reader should select the "entry channel" of his or her choice into approximate methods. For example, mathematical programming can be approached from the area of linear algebra (Fredholm's alternative of linear algebra, Farkas' theorem), or from the area of analysis (calculus, Khun-Tucker conditions, etc.). This may seem a "small" detail but has, nonetheless, caused us to suffer unnecessary delays in our own learning.

The texts below have been ranked according to difficulty. The advanced reader should start toward the bottom of each category. See the bibliography for complete information on the books below.

## Linear Algebra

- Lipschutz (1991)
- Lipschutz (1989)
- Shilov (1977)

- Barnett (1992)
- Pettofrezzo (1966)
- Kolmogorov and Fomin (1970)
- Rockafellar (1970)

## Modeling and mathematics

- Courant, Robbins, and Stewart (1996)
- Ewald (1971)
- Bronshtein, and Semendyayev (1985)
- Lipschutz and Lipsom (1997)
- Langer (1967)

## Economic modeling

- Chiang (1984)
- Downing and Clark (1988)
- Luenberger (1984)
- Samuelson (1983)

# 2

# APPROXIMATE EQUATIONS SYSTEMS

This chapter studies approximate equations systems (AES): what they are, how they modify our understanding of simultaneous equations systems, and how to solve them. This study will enable us to further determine their specific potential in economics and finance.

The first part of this chapter examines approximate equations systems (AES), the second the resolution of such systems, and the third discusses their applications to economics and finance.

AES are not easy to handle. Even small-order systems were very difficult to understand from the traditional calculus optic, and a simple method to solve these systems was not available until very recently. I have developed such a method, which uses combinatorial optimization principles and can be implemented with spreadsheets, which are the main calculation device in modern business practice. The organization of this chapter stresses the critical importance of having a convenient method to solve AES in order to generate practical applications.

A technical note is appropriate at this point. In general, the analysis in this chapter needs only to assume we are dealing with affine sets in the field of real numbers, as in Rockafellar (1970). $\Re$ represents the real number system, which includes integers, rational, and irrational numbers. $\Re^n$ is the usual vector of n-tuples, $x = \{ x_1, x_2, \ldots , x_n \}$. In $A x = b$, the capital letter "A" denotes a real matrix of "m" rows (equations) and "n" columns (variables). It also denotes the corresponding linear transformation: $x \rightarrow A x$, from $\Re^m$ into $\Re^n$. In this monograph, A will be a square matrix of order $k = m = n$, which means there is a different equation for each unknown variable. The simultaneous equations systems used in this book are rather general and undemanding. More information about the mathematical objects we are using and types of extensions they permit can be found in Shilov (1997), and Kolmogorov and Fomin (1970). The former stresses algebra, the latter calculus and analysis.

The first chapter noted that equations were a particular form of relation. One would think such relations would always be imprecise and inexact, except in the most trivial cases and, therefore, that from the very beginning of computing room for error has been included in our specification of relationships. However, this has not been the case. We can not exhaustively enumerate the reasons for not doing so, but the sheer difficulty of including such an error would be among them. Let us start by examining the problems caused by change in a simple SES system.

## 1. CHANGE IN A SES MODEL

Let us, for example, study the two-equation model presented in the previous chapter:

$$(1) \qquad \begin{vmatrix} 1 & -2 \\ 1 & 2 \end{vmatrix} \quad \begin{vmatrix} x_1 \\ x_2 \end{vmatrix} = \begin{vmatrix} -4 \\ 8 \end{vmatrix}$$

or, in algebraic form,

$$(2) \qquad \begin{vmatrix} a_{11} & a_{12} \\ a_{21} & a_{22} \end{vmatrix} \quad \begin{vmatrix} x_1 \\ x_2 \end{vmatrix} = \begin{vmatrix} b_1 \\ b_2 \end{vmatrix}$$

The solution values are $x^* = \{ x_1 = 2, x_2 = 3 \}$. We are familiar with what would happen in the following cases: (1) changes in one of the coefficients aij, or bi, one at a time, and (2) changes in several of the $a_{ij}$'s or $b_i$'s at the same time. The first of these cases is commonly known as **sensitivity analysis**, and is used to determine, for example, how changes in price may affect demand, or the effects of changes in the domestic interest rate on exchange rates. The problem is that even the simple world of the model may not remain constant while a critical variable changes. If the variables are synchronic, that is, if they are related in the same space-time as the idea of simultaneity implies, changes will also be simultaneous. This is why we enroll the help of the so-called **scenario analysis**, where we change several values at once.

We know there are problems in our scenario methodology, but attaching sensitivity and scenario analysis to a simultaneously determined set of causally connected variables represents the state of the art in economics and business practice. Most firms do not get that far into modeling, and it is frightening to think that many critical economic policy decisions—interest rates, budget, currency devaluation, and so on—have not had the benefit of even a simple two-equations model such as the one in equation (1). At the same time, mathematical texts are filled with thousands of procedures that go beyond sensitivity and scenario analysis. They are rarely used in practice and seem to reside in a world of their own where even researchers arrive only after a perilous passage.

What are the problems with scenario analysis? What is next for modeling? How advanced are AES? The answer to the first question is this: (1) changing many things at once does not mean we are capturing what changes, that is, we may be neglecting interdependence, and (2) when several things change in the model, the model itself may have changed. We call this event structural change, and it generates the dreaded "specification problem" of econometrics, which is handled by comparing the original to two (or more) alternative specifications; one of the alternative specifications is said to be "nested" in the original, for example, adding one more equation, and the other is a "non-nested" alternative, for example, specifying the problem in logarithmic (growth) form.

If we assume the change in variables does not affect the basic structure of relationships in the model—and this is a risky assumption—we can use **perturbation analysis** to assess the effects of multiple changes in our model.

Let $A\,x = b$ represent a k-dimensional simultaneous equations system (SES) in the multidimensional space $\Re^k$. Further, let symbol $\Delta$, delta, represent the increment in the values of variables.

We can write, under the usual conditions and assumptions,

(3)  $$(A + \Delta A)(x + \Delta x) = b + \Delta b$$

or

(4)  $$A\,x + A\,\Delta x + \Delta A\,x + \Delta A\,\Delta x = b + \Delta b$$

Assuming the system was at its solution values $x^*$, such that $A\,x^* = b$, and assuming that the increments are so small that second order effects can be neglected, we obtain

(5)  $$A\,\Delta x + \Delta A\,x^* = b + \Delta b$$

and

(6)  $$A\,\Delta x = b + \Delta b - \Delta A\,x^*$$

from where we obtain the potential changes in x, our dependent variables, when both external variables and parameters change around the solution values $x^*$. That is,

(7)  $$\Delta x = A^{-1}\,\Delta b - A^{-1}\,\Delta A\,x^*$$

Equation (7) appears in many books as an approximation to study changes in $x_i$'s, but it is not correct when second-order effects are not negligible. When we do not discard second-order effects, we obtain the following formula:

(8)  $$\Delta x = (A+\Delta A)^{-1}\,\Delta b - (A+\Delta A)^{-1}\,\Delta A\,x^*$$

This formula helps to study changes in $x_i$'s. For example, if b1 increases by 0.5, we would obtain the following values for $x^*$: $x_1 = 2.25$, $x_2 = 2.875$. If both $b_1$ and $b_2$ increase by 0.5, we would obtain $x_1 = 2.5$, $x_2 = 3$. Finally, if we let

$$\Delta A = \begin{vmatrix} 0.5 & 0.5 \\ 0.5 & 0.5 \end{vmatrix} \quad \Delta b = \begin{vmatrix} 0 \\ 0.5 \end{vmatrix}$$

we would obtain $x_1 = 0.45833$, and $x_2 = 3.125$.

If we want to understand how we can improve modeling with AES, it is important that we understand what these procedures mean. The contextual interpretation of the simultaneous equation system described in Exhibit 1.5 indicated the variables in vector b are exogenous, that is, beyond the control of the subject doing the modeling. Changes in these exogenous variables call for changes in the endogenous variables, $x_i$, if there is no change in behavior ($\Delta A = 0$, means $\Delta a_{ij} = 0$ for all $a_{ij}$).

Sometimes changes in behavior are possible. For example, if $b_1$ would also increase by 0.5 in our last calculation, the new values for $x_{ij}$ would be: $x_1 = 0.66667$, $x_2 = 3$. Note that in some cases the value for a given variable ($x_2$ in our example) may not change in spite of the flurry of changes taking place in the system. In those cases we can observe what is known as compensatory changes, which indicate that behavior changed when the situation changed. The increase in oil prices during the 1970s may have provided an example of this situation. Economists and policy makers were extremely concerned because, given consumption patterns in oil, the world would not be able to take the extremely large increases in the price of oil without serious disturbances—disturbances of the type that historically have led to revolutions and wars. The world demand for oil, however, changed and it changed structurally. This changed the "oil picture" dramatically.

Sometimes, however, adjustments are not possible. For example, if a country is already experiencing famine one should not expect much reduction in food consumption patterns, other than by the death of its people. These examples are realistic; they look extreme because very often textbooks have removed issues that inconvenience us. To give a further example, consider how many times we hear that developing countries should "tighten their belts" to solve some of their crises. We do not seem to use fundamental modeling nor basic reasoning in serious situations. (Perhaps, using strange logic is a characteristic of our being human.) Anatole France gave a famous example of the type of logic we seem to favor: "It is only the poor who are forbidden to beg."

During the oil crisis, some countries tried to stimulate aggregate demand—demand management—to stave off recession. However, the increase in oil prices, coupled with increases in raw materials, actually caused changes in the aggregate supply curve. The results were nasty episodes of unemployment accompanied by inflation. This situation could illustrate a case in which a change in one of the bi's (limit supply or limit demand) could have been misidentified or treated with the wrong remedy.

An important idea highlighted in the previous paragraphs is that of solution values for x in A x = b representing an equilibrium between the internal and external worlds of the system. In this context, action in the external ($\Delta b$) or internal ($\Delta$A) environments takes place and must be met by a compensatory reaction ($\Delta$x) in the adjustment variables to prevent the system from collapsing.

SES, as we see, are very important. As we noted in the previous chapter, they represent a structure of intelligibility, which means they are a tool we use to understand the world. Another major reason for their use is the flexibility in interpretations they allow. When we consider these two reasons jointly, we realize SES implement very natural processes in our thinking. Take, for example, the ideas of an input-output relationship, cause and effect, internal and external worlds. They all use SES in some way. SES resemble the ring or structure that balances the forces inside and outside a given situation—its homeostasis. Such a barrier can be solid like that of a castle wall, or fragile like an eggshell. It has measurable strength, both in numerical and informational terms, as we will see next. This is why we may find it advantageous to enrich practical decision-making along the SES-AES pathway.

Coming back to perturbation theory, in the transition from equation (4) to equation (7), we made an assumption regarding the size of the increments that we flatly violated later on. We assumed we would keep our increments very small, so that second-order changes could be neglected. This is an example of abuse of mathematical objects. That we can do something does not imply that what we do makes sense. We may be performing the equivalent of computer-testing how materials in common cars react to light speeds. We can simulate this situation with computers, but we'll see neither light speeds on highways nor common cars in space. However, we still need to study the types of changes, or the size of stress, our structure can withstand. What we need, then, beyond sensitivity, scenario, and perturbation analysis, is a procedure to test the strength of our structure.

## 2. APPROXIMATE EQUATIONS SYSTEMS

When we add error to a SES we obtain an approximate simultaneous equations system. For example, a two-variable system may be expressed in different forms:

(9)
$$\begin{vmatrix} [\,a_{11}^{\,L}, a_{11}^{\,U}\,] & [\,a_{12}^{\,L}, a_{12}^{\,U}\,] \\ [\,a_{21}^{\,L}, a_{21}^{\,U}\,] & [\,a_{22}^{\,L}, a_{22}^{\,U}\,] \end{vmatrix} \begin{vmatrix} x1 \\ x2 \end{vmatrix} = \begin{vmatrix} [\,b_1^{\,L}, b_1^{\,U}\,] \\ [\,b_2^{\,L}, b_2^{\,U}\,] \end{vmatrix}$$

or

(10)
$$[\,a_{11} \pm \varepsilon\,]\,x1 + [\,a_{12} \pm \varepsilon\,]\,x2 = [\,b_1 \pm \varepsilon\,]$$
$$[\,a_{21} \pm \varepsilon\,]\,x1 + [\,a_{22} \pm \varepsilon\,]\,x2 = [\,b_2 \pm \varepsilon\,]$$

where superscripts L and U stand for lower and upper, respectively, in (1), and $\varepsilon$ is a real valued number in (2). Some authors use L (left) and R (right) for lower and upper bounds, respectively.

These alternate expressions explain the dual origins of the field. Interest in approximate equations can be traced to the work of Oettli (1965), Oettli, Prager and Wilkinson (1965), and Kuperman (1971). Systems of interval equations can be traced back to the research of Hansen (1965, 1967), Moore (1966, 1979) in the 1960s, and more recently to Alefeld and Herzberger (1983), Deif (1986), and Neumaier (1990). Research on interval arithmetic spanned further research into functional metric spaces, optimization, reliable computing, and the general optimization (global, nonlinear) problem, see Hammer et al. (1995) and Hansen (1992), respectively.

In general, we can express approximate or interval systems of equations as

$$(11) \qquad A^I \, x = b^I$$

Exhibit 2.1 shows what they look like. The example has been taken from Kuperman (1971) and is the same one we have been using in the SES case. More examples will be provided later. The single point in our two-dimensional example in equation (1) becomes a surface. In a three-dimensional case, the point would have become a cube. In general, we obtain a body, instead of a point.

There are two pictures in Exhibit 2.1, which correspond to errors of two sizes, $\varepsilon = 0.1$, and $\varepsilon = 0.7$, respectively. It is important to note how the solution set changes shape as the error changes. Note also that the errors are the same for each coefficient in A and b. For two main reasons, this is an important assumption maintained throughout this book.

The first is numerical: under this assumption, it can be shown that the system will not be critically ill-conditioned if $\phi$, the critical ill-conditioning factor, is lower than one. If the system is not critically ill-conditioned, it admits solutions. We may introduce what is called the centered form of an AES. This is the simple SES model in equation (1), the one with no errors, which we will designate as $A^C$. The fact that it has a solution is already very significant. In this case the system is not critically ill-conditioned, although it may be non-convex as in the case of $\varepsilon = 0.7$. We can deduct that solutions may exist for a critically ill-conditioned system, but disjointed sets will form.

The second reason is methodological and is also very important. We have been extremely careful not to limit our errors to those arising from measurement or to other errors with probabilistic meaning. We are trying to capture the error that signals overall model reliability, rather than that of a particular parameter, variable, or relationship. We focus on informational strength and numerical, rather than probabilistic, uncertainty. It is also possible that one or more parameters have no errors in an a priori analysis, but regardless we model them with errors. We do this because errors in a given area "infect" other areas of the problem; further, we prefer to make the possibility of a "contagion" explicit.

**Exhibit 2.1**
**Example of Approximate Equations Systems**

$[1 \pm \varepsilon]\, x1 + [-2 \pm \varepsilon]\, x2 = [-4 \pm \varepsilon]$
$[1 \pm \varepsilon]\, x1 + [\,2 \pm \varepsilon]\, x2 = [\;8 \pm \varepsilon]$

<u>Intervals for $\varepsilon = 0.1$</u>                      <u>Intervals for $\varepsilon = 0.7$</u>

$U1 = [\min(x1), \max(x1)] = [\,1.454, 2.667\,]$      $U1 = [-2.666, 16]$
$U2 = [\min(x2), \max(x2)] = [\,2.714, 3.316\,]$      $U2 = [\,1.444, 6.230\,]$

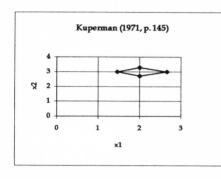

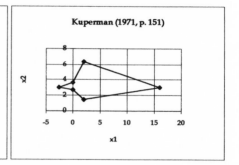

<u>Switching binary-variable model:</u>

|                | (1) | (2) | (3) | (4) |
|----------------|-----|-----|-----|-----|
| $\delta_{11}$  | 1   | 1   | 0   | 1   |
| $\delta_{12}$  | 1   | 0   | 0   | 1   |
| $\delta_{21}$  | 1   | 1   | 1   | 0   |
| $\delta_{22}$  | 1   | 0   | 1   | 0   |
| $s_1$          | 0   | 1   | 1   | 0   |
| $s_2$          | 0   | 1   | 0   | 1   |
| $x_1$          | 16  | -2.666 | 2 | 2 |
| $x_2$          | 3   | 3   | 6.230 | 1.444 |

*Source*: Tarrazo (1998a). © Elsevier Science B.V. Used with permission.

The existence of errors ($\varepsilon \neq 0$) affects both the nature of the solution set and the solutions themselves. First, the solution set is no longer a point but an area in the two-dimensional case, or a body in the three-dimensional case. The existence and shape of this solution set is determined by $\phi$, the critical ill-conditioning factor, as we shall explain later in this section. Second, note that particular solutions for $x_i$'s are no longer single values but intervals or ranges. We shall call these intervals "ranges of uncertainty," to avoid any invalid comparisons to the expression "confidence intervals," which has a completely different (probabilistic rather than numerical) meaning.

We must use AES in economics and business because decisions in these fields of learning are always made under conditions of imperfect knowledge: imperfect, not only because there are many things we do not know about the present, but also because the future is always unknown. The bridge spanning the ground between what we know now and the elusive future is called expectation. Expectations are the heart of economics and finance, and AES provide a way to include them in our models.

Approximate equations are also attractive in economics and finance because the errors may represent missing variables. Let us examine three examples, which have been labeled a, b, and c in the following. In all cases, we could think that our two-dimensional example actually came from three-dimensional cases, with the following values:

$$A \qquad\qquad x \;=\; b$$

a)
$$\begin{vmatrix} 1 & -2 & 0.001 \\ 1 & 2 & 0.001 \\ 0.001 & 0.001 & 5.586074 \end{vmatrix} \quad \begin{vmatrix} x_1 \\ x_2 \\ x_3 \end{vmatrix} = \begin{vmatrix} -4 \\ 8 \\ 0 \end{vmatrix}$$

$$x^* = \{x_1{}^*, x_2{}^*, x_3{}^*\} = \{2.000001, 3, -0.0009\}$$

b)
$$\begin{vmatrix} 1 & -2 & 1 \\ 1 & 2 & 1 \\ 1 & 1 & 6902.594 \end{vmatrix} \quad \begin{vmatrix} x_1 \\ x_2 \\ x_3 \end{vmatrix} = \begin{vmatrix} -4 \\ 8 \\ 0 \end{vmatrix}$$

$$x^* = \{ x_1{}^*, x_2{}^*, x_3{}^*\} = \{2.000724, 3, -0.00072\}$$

c)
$$\begin{vmatrix} 1 & -2 & 0.001 \\ 1 & 2 & 0.001 \\ 0.001 & 0.001 & 0.001 \end{vmatrix} \quad \begin{vmatrix} x_1 \\ x_2 \\ x_3 \end{vmatrix} = \begin{vmatrix} -4 \\ 8 \\ 0.005 \end{vmatrix}$$

$$x^* = \{ x_1{}^*, x_2{}^*, x_3{}^*\} = \{2, 3, 0\}$$

We could say we discarded the third variable $x_3$, because we were mainly interested in the other two, and maybe no harm was done with this assumption. However, the validity of doing so depends on very specific circumstances regarding the values of coefficients in A and equivalent compensatory coefficients in b. If these conditions are not met, our omission of $x_3$ could have definite effects on the values of $x_1$ and $x_2$. The uncertainty concerning the true values of $x_1$ and $x_2$ may be indicated by the size of the ranges built around their centered approximations.

Two fundamental problems in equation (3) are finding a solution $x^0$ (or the smallest box containing a set of feasible solutions), and establishing the range for the admissible solutions, or intervals of uncertainty, for each of the variables involved. Research on reliable computing has stressed the former, while approximate equations research and this monograph focus on the latter. Both of these lines of research are important. For example, they:

1. Recognize the approximate nature of our observations, especially in the social sciences.
2. Account for potential numerical errors.
3. Create room for nonlinearities in our specifications. (For example, a box is a space defined by two interval equations and allows for linear and non-linear behavior within its interior.)
4. Permit better modeling of expectations and uncertainty.

In sum, we started with a point, one that was solitary in a cold space without shadows or light. Then matter ($\delta$) was added and the point became a body, with perhaps glorious Rubenesque proportions. We will explore how this could occur in the second part of this chapter.

We can add something else about bodies. AES, as a form of intelligibility, are bodies in spaces defined within our minds. They are still representations of the masses of information we perceive in the world outside our heads—to use Schopenhauer's terms. We must keep in mind that a phenomenon itself may not be well defined, or that we may not have the cognitive skills (perception) to distinguish it well enough to build an equal (homomorphic), (or at least similar enough) representation of it. Furthermore, the level of resolution we opt for also impinges on our ability to extract a picture from reality. Some problems seem to be better studied at the aggregate (macro) level while others may require disaggregate (micro) focus. We know little about how these issues of **resolution** and **distinguishability** affect the quality of our models.

### A. Approximate Equations and Interval Mathematics

I first learned about interval mathematics (IM) by reading about fuzzy sets, in a footnote situated at the end of a chapter on fuzzy numbers, in Dubois and Prade (1980). Consultation of references in linear algebra, optimization, and calculus found nothing on interval mathematics. This appears to be due not to the lack of importance of the topic but to both the specialization and the novelty of the field. A short reference to interval matrices outside their specialized field appears in Barnett (1992, p. 406).

Our appreciation of the field of IM started by reading the texts by Hansen (1965), Hansen and Smitt (1967), and Moore (1966), both of which reflected a stimulating inclination for practical research. From those texts, I tackled Alefeld and Herzberger (1983) and Neumaier (1990), which are evidence of the maturity reached by the field. My progress was aided by the study of the clearly written study by Deif (1986), which I would recommend as an introduction into the area of interval mathematics, as much as I recommend Kuperman (1971) to anyone interested in approximate equations. Neumaier (1990) offers a wealth of solidly developed procedures and methods that will become instrumental in enhancing our modeling practices in the near future.

AES are not inferior to those IM methods that emphasize precision: They are simply different tools that are better suited for other purposes. AES are like the athlete who instead of specializing in the pure speed of 100 or 200 meter races competes in the decathlon. What he loses in specialization, he gains in his range of skills. As we shall see, in economics and finance the problems may not be very well defined to begin with. This situation calls for a decathlon-type researcher.

Our short account of how we became acquainted with AES perhaps reveals that researchers across sciences and disciplines are looking for solutions to similar problems and that their methods are less different than they once were. Flexibility is added to exact specifications in social sciences as much as engineers seek precision in humanistic systems.

## 3. RESOLUTION OF APPROXIMATE EQUATIONS SYSTEMS

Solving the AES in equation (3) can be easy, difficult, very difficult, or simply impossible, see Neumaier (1990), Kuperman (1971), and Deif (1986). The good news is that some of the cases that may have the most practical value in finance and economics are not those that are very difficult to solve. We hinted at that in the previous section. Furthermore, in those practical cases one can find uncertainty ranges with the help of the familiar, also called degenerate, or centered solution to the AES system, that is, equation (1) in the first section of the chapter.

We have developed a procedure (see Tarrazo, 1998a), which uses an integer, binary, programming (IP) approach to finding ranges of uncertainty in systems of approximate equations like those in equation (12), where each parameter shares the same error. The procedure reformulates the original AES problem in terms of a set of switching variables and the application of well-known IP techniques to the auxiliary problem.

In this section we clarify some of the development of the method. The reader interested in technical details is referred to my previously mentioned article.

### A. The Problems

The two fundamental problems in equations (9) or (11) are to find a solution x0 (or the smallest box containing a solution), and to establish the range for the admissible solutions, or intervals of uncertainty, for each of the variables involved. These have been approached in different ways such as (1) interval arithmetic, (2) determination of the necessary and sufficient conditions for admissible solutions (NASCAS), and (3) linear programming. The applicability of these methods depends on the size of errors ($\varepsilon$), whether $\varepsilon_{ij} = \varepsilon$ as in equation (10), and on the existence of zeros in the uncertainty intervals for $x_i*$'s. For example, many of the methods based on interval arithmetic work best when applied to small intervals, which may not be the case in finance and economics. We found them rather involved as well. Perturbation theory methods can be properly applied only to infinitesimal changes, see Deif (1986), or Kuperman (1971).

In Tarrazo (1998a) we examined approaches (2) and (3) using the example from Kuperman (1971, p. 45), which is also used in this chapter. I cannot make any comments concerning alternative AES solution methods without a few observations concerning the solution set of approximate systems. These observations will add to those made in the previous section and will provide further clues concerning dimensionality and computational complexity.

Let equation (1) be the centered form of the following AES systems, which is Kuperman's example:

(12)        $[\,1 \pm \varepsilon\,]\, x_1 + [\,-2 \pm \varepsilon\,]\, x_2 = [\,-4 \pm \varepsilon\,]$
             $[\,1 \pm \varepsilon\,]\, x_1 + [\,\ 2 \pm \varepsilon\,]\, x_2 = [\,\ \ 8 \pm \varepsilon\,]$

In order to solve this system, that is, to find the uncertainty intervals for the variables involved, we must first ascertain that there is no critical ill-conditioning. Doing so is substantially more difficult in the case of interval or approximate equations than for the noninterval (or degenerate) case. It is easier to assess critical ill-conditioning under the assumption that $\varepsilon_{ij} = \varepsilon$ in equation (2), that is, if the errors are of the same magnitude for each of the coefficients, see Kuperman (1971). This is an important assumption maintained throughout this monograph and which is well suited for practical cases in finance and economics, where we may not know enough to assume otherwise. Under this assumption, it can be shown that the system will not be critically ill-conditioned if $\phi$, the critical ill-conditioning factor, is less than one. Solutions may exist for a critically ill-conditioned system, but those will form disjointed sets.

Let $A^C$ represent the "centered form" of the approximate system, and let Z be its inverse, with zij as the typical element. The typical element of $A^C$ would be $a_{ij}$, $a_{ij} = (\,a_{ij}^L + a_{ij}^U\,)/2$, in equation (9), or equation (10) with $\varepsilon = 0$. Then, the critical ill-conditioning factor, $\phi$, can be calculated as follows.

(13)        $\phi = \sum_{i=1,k} \sum_{j=1,k} \mathrm{abs}\,(z_{ij})\ \varepsilon$

The centered form, equation (1) or equation (12) with $\varepsilon = 0$, has a solution of $x^0 = [2, 3]$, and a critical ill-conditioning factor $\phi = 0.15 < 1$, which indicates a nondisjointed, possibly convex set. Looking at Exhibit 2.1, we see that with $\varepsilon = 0.1$ we do have a convex set, but with an error of $\varepsilon = 0.7$ the solution set is not convex: it is not possible to go from one extreme solution to the next without "leaving" the solution set.

The centered form provides information on two critical areas: (1) ill-conditioning, and (2) dimensionality of the solution. If an approximate system of equations has at least one solution, say $A^C$, then it is not critically ill-conditioned and, again, the set of admissible solutions forms an insular set—that is, it is not disjointed. Note that the factor $\phi$ lets us determine the sensitivity of the system to the values of $\varepsilon$, which we will use in our applied cases in the following chapters as an indicator for the error-carrying capacity of the structure of the system, that is, the matrix A in A x = b.

Barring critical ill conditioning, the solution set to an AES is a polytope, or bounded polyhedron, in the Euclidean $\Re^k$ space. The NASCAS approach amounts to finding SES subsystems within the structure of the problem that will provide extreme values for each of the $x_i$'s. In other words, we use SES subsystems to find the extrema or corner points of the polytope defined by the structure of reference. Since we need k equations to establish an absolute reference (or point) in the $\Re^k$ Euclidean space, we obtain each of these from the NASCAS. Moreover, in $\Re^k$ we need $2^k$ points to define a full dimensional (bounded) body. We know that a polyhedron P is full dimensional if dim (P) = k, and if and only if it has an interior point. The polytope in our example of Exhibit 2.1 full dimensional because the solutions to the centered form $A^C$, with dimension k, provide such an interior point.

This search of extrema for $x_i$* using NASCAS is carried out with a great amount of logic, as our study and Kuperman (1971) show, and is an excellent activity to learn about solving AES. While it is possible to write a program for specific, low-order cases, it can be prohibitive to do so for higher-order cases. However, in general, the simultaneous equations approach applied to the NASCAS becomes intractable if (1) there are more than a few variables, or (2) if the variables have solutions in other than the positive orthant. Linear programming helps in the first case, but not the second.

Applying linear programming (LP) amounts to performing [2*k] optimizations, one per bound and per variable, subject to the NASCAS, which implies that we still have to lay them out to be able to use linear programming. If all the variables fall in the positive orthant, LP is useful. If, however, the solution set lies in more than one orthant we must determine the constraints that are relevant for those orthants: daunting task because the set of constraints obtainable from the NASCAS is different for each of the 2K orthants of the solution space. There is some good news: One way to find whether the interval equations fall in more than one orthant is to obtain the solutions for the centered form and observe whether $x_i > 0$ for all i. Kuperman's (1971, p. 151) example ($\varepsilon$ = 0.7) illustrates this point. But there is worse news: The set of all feasible solutions may not form a convex set and, very unfortunately, nothing can be said about potential nonconvexities unless all intervals of uncertainty are calculated. The right-hand side of the graph in Exhibit 2.1 shows that the solution set is not convex; therefore, relaxing $\varepsilon$ the non-negativity conditions will not provide us with the whole set of solutions—the simplex algorithm cannot jump from one extreme point to another. For non-critically ill-conditioned systems, nonconvexities may be spotted by either visual inspection of the graphics (which is useful only for systems of a couple of equations), or by observing the presence of zeros in the interval of uncertainty for each of the variables involved. These means that we can put a lot of time and effort into trying to obtain solutions for AES systems using LP methods and come back empty-handed.

As Kuperman (1971, p. 187) notes, the only available approach (including interval mathematics methods) in this case is to use the NASCAS directly, which is burdensome even for a two-variable case.

### B. A Way to Solve AES

As we note in Tarrazo (1998a), we propose reformulating the problem of finding ranges of uncertainty for AES as a switching binary-variable model, as in the following manner. Let us define a set of variables vij, ti

(14)  $$v_{ij} = a_{ij}^{L} + \delta_{ij} \; (a_{ij}^{U} - a_{ij}^{L})$$
$$t_i = b_i^{L} + s_i \; (b_i^{U} - b_i^{L})$$
$\delta_{ij}$ and $s_i$ are binary variables

Then perform the following optimization to obtain the upper (maximization) and lower (minimization) bounds for variable xi.

(15)        Select            $\delta_{ij}$ and $s_i$
            to maximize      xi (minimize if lower bound preferred)
            subject to       $V x = t$, where $v_{ij}$ and $t_i$ are the typical
                             elements in V and t.
            $\delta_{ij}$ and $s_i$ are binary for all i, and ij.

The model specified in equations (14)–(15) finds the correct uncertainty intervals for Kuperman's $\varepsilon = 0.7$ example, (1971, p. 151); Exhibit 2.1 shows those correct uncertainty intervals and the corresponding values for the binary variables. Columns (1), (2), (3), and (4) in Exhibits 2.1 and 2.2 represent the cases of maximizing $x_1$, minimizing $x_1$, maximizing $x_2$, and minimizing $x_2$, respectively.

Exhibit 2.2 presents the solutions to a nonconvex example given by Hansen, which is reproduced in (1992, p. 26) and studied by Deif (1986).

Our procedure combines the strengths of classical (SES, NASCAS) and operative (programming) methods. Programming methods are very good for searching, classical methods are good for uncovering structures and providing the solution values implicit in those structures. Our procedure uses the SES subsystems, which the switching variables use to find the extrema or corner points of the polytope. The binary variables ensure that only extreme values will be used in each search. The search carried out with the help of the binary variables is efficient and takes advantage of the raw power of our current hardware for exhaustive numerical computation; the procedure selects the best constraints to perform selective enumeration.

**Exhibit 2.2**
**Nonconvex Example of AES**

Example in Hansen (1992, p. 26), Deif (1986, p. 51)

$[2, 3] x_1 + [0, 1] x_2 = [\ 0, 120]$
$[1, 2] x_1 + [2, 3] x_2 = [60, 240]$

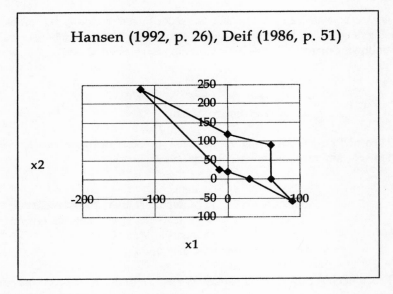

|          | (1)  | (2)   | (3)   | (4)  |
|----------|------|-------|-------|------|
| $\delta_{11}$ | 1    | 1     | 1     | 1    |
| $\delta_{12}$ | 0    | 0     | 0     | 0    |
| $\delta_{21}$ | 0    | 0     | 0     | 0    |
| $\delta_{22}$ | 1    | 1     | 1     | 1    |
| s1       | 0    | 1     | 1     | 0    |
| s2       | 1    | 0     | 0     | 1    |
| $x_1$    | 90   | -120  | -120  | 90   |
| $x_2$    | -60  | 240   | 240   | -60  |

Note: Columns (1), (2), (3), and (4) represent the cases of maximizing x1, minimizing x1, maximizing x2, and minimizing x2, respectively.

In general, enumerative techniques take advantage of the fact that in a bounded (linear) IP problem the set of values of the integer variables is finite. Furthermore, the number of critical variables in certain areas of inquiry such as economics and finance may be small enough for the researcher to contemplate exploring combinatorial optimization techniques. In these cases, one could implement selective enumeration, see Spielberg (1979) in the preprocessing stage of the problem, see Nemhauser and Wolsey (1988).

Note that binary programming (selecting extreme values) represents a generalization strategy to deal with the complexity caused by the combinatorial structure of the problem.

In sum, the procedure we propose finds the set of feasible solutions and the ranges of uncertainty for $x_i$'s even in nonconvex cases. It also lets the computer do all the work. Both the linear programming and NASCAS approaches relied on the researcher to provide the constraints that were relevant for each of the instances of the problem solution; the binary formulation of the problem lets the computer do the work for us.

### C. A Comment on Computational Efficiency

Reformulation of the interval equations in terms of switching variables finds the correct intervals of uncertainty for the $k=2$ case very efficiently. What about higher-order problems?

Our early experiments with AES were carried out on what was, at the time, a very average desktop computer (486DX266, 16MB RAM) and a popular spreadsheet program (Microsoft Excel 5.0 for Windows 3.1). We had access to better equipment but felt a practical method should be good enough to be implemented under average conditions.

Using this very modest equipment, we generated a 10x10-integer coefficient matrix and a 10x1 vector from a normal distribution. The error ($\epsilon$) was set to 0.1—data are available from the author. This means the program had to optimize 110 binary variables in the calculation of every corner of the polyhedron. The centered form AC had an inverse, which indicates the set of all feasible solutions to be insular ($\phi = 0.43 < 1$). Variables $x_1$, $x_5$, $x_8$, $x_9$, and $x_{10}$ had negative optimal values for the centered form, meaning the set of all feasible solutions spread beyond the positive orthant. Note again: nothing can be said about potential nonconvexities unless all intervals of uncertainty are calculated. Excel's optimizer, the Solver, was set with the following options:

| | |
|---|---|
| Max time: | 1,000 seconds |
| Max iterations: | 1,000 |
| Precision: | 0.000001 |
| Tolerance: | 5% |
| Estimates: | Tangent |
| Derivatives: | Forward |
| Search: | Newton |

Under these conditions, it took the optimizer about 4 minutes (48 trials) to find the maximum value for $x_1$=-0.097709, and about 7 minutes (55 trials) to find the minimum value for $x_1$= -0.573218. These results were obtained by setting all the initial values for the binary variables to one after each optimization. When the minimization was started from the maximization case, it took the program about 13 minutes (108 trials) to find the minimum value for $x_1$, which gives us some information concerning the time it took the program to travel from one extreme value to another in the polytope. It may take about 80 minutes (4 minutes x 20 optimizations) for this computer to solve a $k = 10$ case—a very, very small amount of time and effort compared to the prospect of deriving the relevant constraints for each possible orthant by hand, which can take several days of intensive work for a ten variable case. It was obvious that, with our procedure, professional computers with specialized software should be able to handle with ease cases of about a 100 variables.

Our findings were particularly encouraging because we used only general purpose, branch-and-bound methods for IP programming, which obtain their 0–1 solutions from the fractional solutions. Doing so is highly inefficient compared to using specialized algorithms for binary programming, see Balas (1965) and further improvements by Balas (1967), Glover (1968) and Geoffrion (1967). Note that we find ranges of uncertainty with modest computing power and general-purpose IP methods for cases that become intractable with current approaches. Interval mathematics methods often require highly specialized software—for example, Hammer et al. (1995). In contrast, our approach to finding intervals of uncertainty can be implemented with limited hardware and software resources, as we set out to do.

The convenience of our procedure should encourage the use of approximate equations in both theoretical and applied research as well as in practical endeavors, especially in those situations where our knowledge of the problem permits models to be specified with only a few equations.

The role of approximate equations in practical applications looks even better in light of current developments in computing and the Solver technology included in Excel. The calculations in this monograph were carried out on a Pentium-MMX 266Mhz laptop (64MB of DRAM) of the type commonly used in business endeavors with the usual business software: Microsoft's Excel 97. The Solver technology included in this version of Excel handles binary constraints, which makes it very easy to specify and solve the approximate equations problem. In sum, calculations originally performed on the old machine were carried out much more rapidly on the new, even without dedicated binary programming. (This is the information provided in the Help facility of Microsoft Excel on "Algorithm and Methods Used by Solver": "Microsoft Excel Solver uses the Generalized Reduced Gradient (GRG2) nonlinear optimization code developed by Leon Lasdon, University of Texas at Austin, and Allan Waren, Cleveland State University. Linear and integer problems use the simplex method with bounds on the variables, and the branch-and-bound method, implemented by John Watson and Dan Fylstra, Frontline Systems, Inc. For more information on the internal solution process used by Solver, contact: Frontline Systems, Inc., P.O. Box 4288, Incline Village, NV 89450-4288. (702) 831-0300. Web site: http://www.frontsys.com; Electronic mail: info@frontsys.com.")

Convenience and simplicity were two guidelines we imposed on our search for a practical procedure to solve approximate equations. As we shall stress in the chapters concerning practical applications of approximate equations, we also favor a simplification strategy to deal with complexity. My research and experience in economics and finance taught us that in many important cases we do not know enough to handle models for all but a few equations and a few variables. Each of the following chapters that dwell on applications will provide evidence of that. One implication of this reflection is that there is enough computational capacity in the average home computer to handle practical applications of approximate equations.

Our objective of developing a convenient procedure to solve approximate equations systems in practical settings has been accomplished.

## D. The Search for the Procedure

A short account of how we developed the procedure will serve to highlight the potential of both the strategy employed in the numerical solution as well as the AES themselves.

I first invested some time learning about different methods to solve AES via interval arithmetic and tried to obtain some of the specialized software mentioned in some interval mathematics texts. I was looking for a simple procedure that could be implemented with readily available software and applied to a fairly general case. Kuperman (1971) and Deif (1986) were significant findings in this search, because their presentation of each of those methods was very clear and pointed out comparative analyses of their adequacy and limitations.

First, I selected Kuperman's case with equal size errors in matrices A and b, because it represented a general case and provided a base to proceed further. This base of operations, furthermore, housed the centered form $A^C$, which is the familiar well-known case. Secondly, I opened a spreadsheet file (Microsoft Excel) and studied the problem. Spreadsheets are the most popular computational software in business. They include matrix procedures and a wealth of mathematical operations. As noted, they also include an optimizer called Solver which, in addition to handling general purpose linear and nonlinear optimizations, includes fractional and integer mathematical programming procedures. Another advantage of spreadsheets is that they allow modeling to be implemented in a hierarchical and modular way, and they, themselves, are a fine information management tool. Our experience with spreadsheets, see Tarrazo and Alves (1998a), attests to their general reliability.

We initially tried searching the solution ranges via intermediate values, that is, by averaging the extreme (upper and lower) cases. This did not bear fruit, because averaging provides only single points and there are infinite ways to split an interval.

We then focused on extreme values for two reasons. First, by doing so, we lowered the number of possibilities to test. Second, from a cognitive viewpoint, we realized that what is useful about ranges is not an average value built from upper and lower bounds, but rather in the range itself, or the extremes (what is known as the spread in the stock market). For example, when we take a shower, we care about the extremes of hot and cold that may come out of the faucet—these are the values that can hurt us.

We tried to explore a low-order case analytically, by hand, but it seemed very hard to discern regularities that can be used in systematic searches. Another more serious problem was that of the sheer number of combinations to be tried. We started to think about building an auxiliary problem that would envelope the original, and about using intermediary variables to select upper and lower parameters that could provide the ranges of uncertainty for the $x_i$'s. This strategy implements Leibniz's following remark: "It is unworthy of excellent men to lose hours like slaves in the labor of calculation which could be safely relegated to anyone else if machines were used." (Coveney and Highfield, 1995, p 46.)

The notions of extremes, combinatorial, and intermediary were critical for us to arrive at the switching binary variable method. We first tried setting $\delta_{ij}$'s as follows: $a_{ij} = a^U + \delta_{ij} a^L$. We obtained solutions, but they were unreliable. Later, we settled for the specifications noted earlier:

$$
(16) \qquad
\begin{aligned}
v_{ij} &= a_{ij}^{L} + \delta_{ij} \ (a_{ij}^{U} - a_{ij}^{L}) \\
t_i &= b_i^{L} + s_i \ (b_i^{U} - b_i^{L})
\end{aligned}
$$

$\delta_{ij}$ and $s_i$ are binary variables

Binary auxiliary variables such as $\delta_{ij}$ and $s_i$ seem to be a natural complement to variables expressed as ranges. In general they also seem to capture the most salient features of choice among alternatives that are also expressed as a range of values. For example, shortly after we developed our solution procedure for AES systems, we were able to apply those procedures not only to the calculation of range-optimal weights in a portfolio, Tarrazo (1997f), but also to the case of solving a portfolio under realistic short sales restrictions, Tarrazo and Alves (1998b).

### E. Linearity

The procedure we have presented and that we will use in the following chapters on applications has some important methodological implications. First, the binary variable approach that enables $x_i$'s to be handled as switching variables represents a generalization strategy to deal with the combinatorial optimization nature of the problem, as was noted earlier. But it also handles the problem of nonlinearity in a special way. The method travels from the heart of the problem, the centered form, to the extreme points of the polytope—the linear body enclosing the $x_i$'s. In doing so, it allows for nonlinear behavior on the inside of the polytope, which has important theoretical implications as well.

As mentioned earlier, considering ranges in a SES problem turned a single point into a body, "with perhaps Rubenesque proportions." Admittedly, these objects look more cubist, as if thought out by Picasso, rather than Rubens. Cubist representations are often linear with angular borders, while Rubens' objects are round, nonlinear, and richly textured. This was not, of course, a matter of our choice but a consequence of the type of objects in the analysis: affine and linearly convex spaces. Linearizations describe crystal-like structures, which derive their piercing power of expression from the simplicity of their forms and can be taken as a (mathematical) simplification of rounded forms, very much as an octagon outlines a circle by either inner or outer linearization. As the facets, linear or plane cuts, in a polyhedron increase, the figure resembles that of a sphere more and more.

One may argue that the divide between linear and nonlinear structures may be more important in aesthetics than in mathematics. Note that the optimization of a multidimensional nonlinear body always rests on finding a saddle-point, or a separating hyperplane between two nonlinear structures, which is itself linear. Thus, linear SES are, if not both the first and last object we encounter in optimization, at least the final objects we handle when we optimize nonlinear structures in static or dynamic terms—see Samuelson (1983), Luenberger (1984). This is even more true when dealing with integer variables, see Nemhauser and Wolsey (1988). Moreover, in classical optimization we often look for a point in $\mathfrak{R}^k$, which is the absolute optimum—assuming it exists—in the dimensional space of the problem. Furthermore, we know that a single point can always be represented as a solution of a SES. No matter what is said in favor of linear models, nonlinear options are the north we look at in our modeling. Nonlinear specifications allow us to model self-adjustment properties in a given system, and allow for realistic dynamic behavior. Chapter 6 will revisit this issue.

In any case, AES deal with bodies or boxes, and within each box both linear and nonlinear behaviors are possible. Therefore, we can say that when we use AES we do not discard nonlinear behavior, nor its dynamic and system implications.

## 4. APPLYING AES IN ECONOMICS AND FINANCE

We are now in a position to more reliably assess the use of AES in economics and finance.

In addition to the observations made at the end of the first chapter, we may note the following in favor of modeling with AES:

1.   It relaxes the assumption of "completeness." The errors in the AES model may come from missing or insufficient variables, as seen earlier in this chapter.

2.  It accounts for errors in measurement and in the definition of the variables, which are so common in economics and finance.
3.  As noted, AES address the problems of inaccuracies in the specification, or the detection, of functional relationships. In one of his Fragments, Heraclitus—the master of change—noted, "a hidden connexion is stronger than an obvious one."
4.  Because of 1) and 3) above, AES address the general misspecification or model selection problem. They may even be able to help in choosing static versus dynamic specifications, since a box permits both statics (inflexion points) and dynamics (trajectories) within it.
5.  By construction, AES allow for blurriness in the definitions of external-independent-exogenous-forcing versus internal-dependent-endogenous-forced variables formulations. In many practical cases we know there must be a balance between sets of variables, but classifying them along the previous lines is difficult.
6.  AES offer potential to mimic individual representations. If two agents select different "pieces of reality," or the same reality with different specifications, the models will be different. The errors allow some margin of maneuvering for each model to do its job.

It is appropriate to examine some observations by Caws (1997):

1.  "The trouble is that the series of experiences is different for each subject," whereas science requires entities having "a structure which is the same for all subjects, even though they are based on such immensely different series of experiences." Caws (1997, p. 176)
2.  Caws' optimum complexity principle: "For a given individual in a given culture at a given historical and developmental moment there will, as a condition of mental stability, be an optimum level (or range) of complexity in the structures subject to intentional animation at a given moment, and that this optimum will be constant or will change only slowly with time." (Caws, p. 217)
3.  "As the world becomes more complex, more of its detail will be sacrificed to a process of smoothing over; in the hierarchy of systems we will choose to operate at a level closest to our optimum even if the elements on that level represent great complexity on the next level below." (Caws, p. 217)
4.  "[This is] alarming because there are things we will never do, reassuring because this limitation will oblige us to rethink the structure of each domain of human knowledge in each generation to preserve both its systematic character and its accessibility to instructed individuals in the next generation." (Caws, p. 218)

Caws' first set of statements highlights the fact that building exact replicas, or expecting farmers in a barter economy to use the same representations of financial systems as IMF officials, is irrelevant. Each one of us has different training and cognitive skills; we may face the same problems, and we may solve them in the same way, but each of us may be using a different, individual representation.

Caws' last comment stresses that a valid representation may take the form of a skeletal wire-frame structure, an exact replica, or a metaphor. It is likely we will not be able to build exact replicas and, if we are, we may be dealing with small and inconsequential matters. What is important is the ability of the subject to form representations that help him solve his problem. Even the choice of objects (natural language, simultaneous equations, and poems) may be secondary in determining the usefulness of the representation.

In sum, this is a short account of the reasons to use interval methods in economics and finance:

1. *Numerical.* There are always measurement errors, which are often caused by having to use indicators instead of the variables of interest themselves.

2. *Probabilistic.* The statistical theory we use all over finance is frequency based, which has a hard time dealing with structural change. Economic and business decisions are about the future, a setting in which everything is changing.

3. *Theoretical.* Theoretical models in finance are very rough representations of reality when compared to those of physics or mechanics. Note that we can send a probe to Mars, or a rocket to Jupiter, but we do not have the math to manage a simple portfolio of stocks.

4. *Informational.* The calculus-plus-frequency-based probability objects in prevailing theoretical models cannot incorporate a great deal of relevant quantitative and qualitative information.

5. *Behavioral.* People may make cognitive and behavioral mistakes, which separate the predictions of our "exact" models from observations of facts. Intervals are a way to add tolerance to the exact models, so that they can incorporate "real" behavior. We know little about these potential mistakes, or about how they affect theoretical models.

6. *Cognitive.* Real people have cognitive abilities that are different from those implied by theoretical models. Humans can handle imprecision and nonfrequency-based uncertainty. They do not need to measure with infinitesimal precision when they are considering buying a house, launching a new product, or planning for retirement. Our knowledge representations (and cognitive maps) are schematic but efficient, perhaps because they capture major effects.

These reasons give approximate equations the edge over the accuracy or error control in reliable computing. The correction of errors may be critical in some areas, but in inexact sciences we will always have to live with approximations.

The characteristics of modeling with AES as well as the reasons to use them in economics and finance suggest some guidelines for modeling with AES in these fields. These guidelines are preliminary because each problem-subject pairing may require its own strategy. What seems to be important, however, is to build a set of relationships, an informational structure. That is what makes the world intelligible to us. This set should include the main features of the problem being studied in a logically complete manner. There will be errors, omissions, and inaccuracies, but hopefully they will not be so strong as to invalidate the overall approximate model. When we use this type of model we may be after the quality of relationships rather than the quantity of variables.

AES model building should reflect Zadeh's incompatibility principle as well as Caws' observation of subject's cognition, and others that echo Zadeh's remark.

Preliminarily, at least, AES modeling seems to favor models of a few equations and variables linked by well-known relationships, which stress the most fundamental features of the problem at hand. We believe this is our best chance to preserve the basic principles of modeling (distinguishability, representability, and measurability). Using approximate equations with many variables and functional relationships defeats the rationale, strengths, usefulness, and purpose of the exercise.

We have all surely experienced as a financial manager and as a researcher what Bertrand Russell noted in *The Philosophy of Logical Atomism*: "Everything is vague to a degree you do not realize till you have tried to make it precise" (1985, p. 38). Equipped with AES, however, we are better prepared to tackle practical problems. In economics and finance we have considered hundreds of years of effort trying to show higher t-statistics and R2's, yet we seem to have forgotten the people (and the world itself) in those models. Meanwhile, the world has become more complex and harder to explain, and, correspondingly, individual agents need more support than ever. Therefore, in our applications we will be much more modest in our goals. We simply want to show how AES can be used to build realistic, subject-specific models, which can, in turn, be used to support decision-making.

## SUMMARY

This chapter focused on approximate equations, which were presented not only as a mere extension of simultaneous equations systems but as something with their own, unique characteristics that offer great promise in practical endeavors.

It first reviewed the limitations of conventional SES to handle change, then reviewed approximate equations and their origin. At that point, it became clear that approximate equations were attractive in economics and finance because they allow for errors of a different kind than those usually considered. AES provide an indicator of the overall "informational strength" of the structure in our model. Unfortunately, despite its attractiveness, it also became clear in order for AES to be implemented practically, we would need a simple, convenient, efficient, and theoretically sound technique to solve them. We provided one such technique and, therefore, cleared a major hurdle to implementation of approximate equations.

Our knowledge of AES allowed us to clarify some of the motivations for using them in finance and economics. We also stated some important additional reasons for doing so. One way to sum up the rationale to use AES in finance and economics is this: we must use AES in economics and business because decisions in these fields of learning are always forward-looking and always made in conditions of imperfect knowledge. The bridge that covers the ground between what we know now and the elusive future is called *expectation*. Expectations are the essence of economics and finance, and AES are an operative way to include them in our models.

# 3

# TOWARD PRACTICAL-
# APPROXIMATE-
# MACROECONOMIC MODELS

In this and the following two chapters, we will present applications of approximate equations. This chapter will focus on the economic environment. Chapters 4 and 5 will concentrate on corporate financial planning and portfolio management, respectively. We must also note that in our practical applications we will leave out issues stemming from human interaction, such as organizational behavior and rationality in decision-making. Rather than examining whether individuals are rational or irrational, we focus on the development of useful representations, which we deem to be a more productive avenue for research into individual cognition. Eliminating these human elements is contrary to what happens in practice and is yet another motivation for including errors in our analyses.

It is critical for the purposes of this monograph to understand that we view models as representations. We do not accept academic models as the "only way to go" but only as generally good starting points. Beyond that, we want to make these representations realistic by strengthening their logical consistency regarding the world within the control of the agent (endoconsistency) and the world outside the control of the agent (exoconsistency). Good decisions are always at the border of these two worlds as the expectations (xi's) are in the expression $AI \; xI = bI$. It seems our understanding of the balance between those worlds is what makes for useful planning.

It seems paradoxical, but our experience in business and investments management seems to indicate that while decision makers always take into account the economy—either implicitly or explicitly—in their planning, they may not do it as economists assume, that is, with quantitative macroeconomic or econometric models. We will understand why when we examine macroeconomic theory and econometrics in search for practical macromodels. In a way, the first part of this chapter makes the point that we will all die of old age before conventional macroeconomic modeling comes up with models of practical value. Something different must be tried, and approximate equations may represent a step in the right direction that is so sorely needed in macroeconomic modeling.

## 1. THE PROBLEM, AGENTS, AND MODELS

In this section we are going to review the most salient features of quantitative macroeconomic modeling and why approximate specifications may help a great deal in making theoretical models practical. Simultaneous equations systems play a fundamental—if not the only—role in economic modeling.

First, we review the most fundamental notions of economic modeling. Then we will concentrate on IS-LM models. The last segment in this section notes that economic knowledge goes beyond quantitative macroeconomic modeling. It is possible that what we consider nonquantitative modeling may add to our treatment of systems in approximation form in the future.

### A. Economic Modeling

The word "economics" comes from the Greek "oikos" and "nomos," which make reference to "house" and "administration," respectively. However, economics was born to study how people become wealthy, not in a spiritual, philosophical, or moral sense, but in a pecuniary way. Adam Smith's inquiry into the causes of wealth of nations is taken as the first work in what we understand as modern economics. At that time, however, economics was hard to disentangle from her social science sisters of law, political theory, and sociology.

At some point, economists split their analyses of the economic reality by focusing on either the aggregate or the individual units. In other words, some chose to work with a microscope, others with a telescope.

Microeconomics has been stagnant for a long time and the hundred-year old Neoclassical tenets concerning demand and supply theory are still commonplace in textbooks and lectures (see Henderson and Quandt, 1980; Samuelson, 1983; or Varian, 1984). This neoclassical theory has also led to such a strong standardization of economic agents, so as to render these idealizations useless for any practical purpose. The typical consumer, producer, or market in microeconomics textbooks is a fiction. They are used, however, because investigators had ulterior motives when they built such agents. They foresaw the distinct possibility of adding up the myriad of consumers and producers (Walrasian crowds) into two camps that would meet in the area called the market, and generate what is known as general equilibrium. Mathematical models of general equilibrium also depict complete fictions, which only explain the world inside the heads of some economists. We are not aware they have provided any practical testable hypothesis yet, or that it can assist agents (including governments) in any practical way. In fact, general equilibrium is an excellent example of Zadeh's incompatibility principle: detail obtained at the cost of relevance, and generality at the cost of realism.

The telescope (macroeconomics) is still being used though it has experienced many mutations, some of which concern the following elements:

1.   The amount of politics explicitly included in the model. The idea of turning macroeconomics into an apolitical activity via mathematical positivism has failed.

For the most part, politics and economic policy are indivisible and the old name of "political economy" is the best descriptor of macroeconomics.

2.  The role of government in and out of economic models. The three major questions our society faces are (a) what goods and services are to be produced, (b) how to produce them, and (c) how to pay for the process? The public sector, which is already bloated in terms of expenses and taxes in almost every developed country, is expected to produce public goods and take the lead in energy conservation, education, ecological activity, natural resource management, peacekeeping, social retirement programs, public health care, and so on. There is no question about the desirability of efficient decision-making in those areas, and in some of those areas corporations simply cannot do this. The problem is how to pay for this enlarged role. Macroeconomic models are extremely ill prepared to handle these forward-looking issues.

3.  The nature of the private sector. How efficient is the modern private sector left to its own? We may never know. If we take the private sector to be very efficient, then the stabilizing and "pump priming" role of the public sector in the economy is greatly diminished. This is a difficult matter, which, nonetheless, is at the heart of macroeconomic model building.

4.  Economic expectations. There are two ways in which the private sector can be efficient: (a) price flexibility in reasonably complete multi-period markets, and (b) rational expectations. Including expectations in economic models has been the major development in economic modeling in recent history. Once we include expectations, it makes sense to model them in particular ways. Rational expectations are those that are correct on average, that is, they are not systematically wrong. One of the problems with including expectations is that they considerably complicate the mathematics of model building. For example, they may require dynamic models, expressed as discrete stochastic calculus and linear difference equations, which become rather complicated for even small (three equations, three variables) models, see Begg (1982). To reduce the entire macroeconomic environment to three equations certainly seems presumptuous; yet, certain economists are fond of using such simplistic models to argue on topics of great caliber such as money neutrality, the relationship between inflation and unemployment, private sector stability, or monetary control (see Barro, 1990). What they find is ... "more tests of this kind are needed to settle the issue," or "the evidence on the issue is still fragmentary and somewhat mixed." Sargent (1987, chapter XVII, p. 460). No wonder! Frustration is due both to the attempt to exclude politics from economics and, very especially, to obstinacy in using most exact methods—we include probability here, too, in that it allows for only "tame and well-behaved" exceptions—in an area where everything is approximate, at best.

5.  Another major emphasis has been that of endowing macroeconomic models with "microfoundations." If we had any decent model of choice for consumers or producers, this strategy would make sense. However, the microfoundation effort invariably amounts to simply refrying the same old neoclassical utility and production theory in the oil of dynamic specifications. The problem is that these theories do not describe and cannot be used by real agents (consumers, investors, or firms). Economic models with these microfoundations resemble mirage cities resting on quicksand.

6.  Modeling the whole versus modeling pieces separately. The quest for the "Absolute Model" in macroeconomics is still on; many economists cannot get over

the need for an all-revealing gadget. But much effort is being done to develop realistic models depicting or describing parts of the economic reality, which seems to be a more reasonable goal. Besides, any exposure to economic modeling shows that different problems call for different methods.

In addition to the previous observations, we must entertain the thought that some models may be appropriate for learning about economics, even though they may not have much empirical worth. Finally, economists, policy makers, and business-people have different needs from what an economic model should do for them. Economists may stress the abilities of the model to explain general questions such as the efficiency of the price systems and the effectiveness of economic policies to increase wealth and employment. Businesspeople, however, may prefer models that are tailored to the problems they face on a daily basis.

The reader interested in further evaluation of economic model building is referred to Kuttner (1985), which is a perceptive essay on the ills afflicting economics, to Thurow (1984), which is a refreshing example of economic analysis developed in a discursive, qualitative, political economy way, and to Samuelson (1983), who provides an example of nondogmatic, nonrigid formalism. Samuelson (1983) is an excellent example of how to use different analyses for different purposes, and in using mathematics as a vehicle for learning, instead of using it for building veils with which to disguise ideology to serve our political purposes.

The pervasiveness of simultaneous equations is evident in economic model building. In microeconomics, production theory studies a balance between input-output relations, consumer theory that between earnings and expenses, and market equilibrium the balance between supply and demand via prices—for example, $S(p) = D(p)$. In macroeconomics, static or dynamic formulations of the labor market, aggregate supply-demand relationships, and interest rates-income relationships invariably build upon SES models.

### B. Quantitative Macroeconomic Modeling

We call economics the princess, nay (Hobbes would say), the queen of social sciences because it uses mathematics and economists use impressive prohibitive argot and sophisticated quantitative models. Some sobering facts, which we state frankly and without remorse, should correct our Jungian-like self-inflation:

1.  We do not have any (simple or complicated) quantitative macroeconomic model that can be used for forecasting purposes. Sometimes, especially in cases of severe need (for example, lawsuits), some agents pay for forecasts built by firms specializing in that trade. No one knows how good these models are or how they are "cooked" and, perhaps as the American saying goes, "It is best not to ask how canned meat is prepared."
2.  It is unclear who, if anyone, uses in practice the quantitative models we have or how they are used.

3.  Reliable partial quantitative models of, for example, interest rates, unemployment, exchange rates, and so on, are also hard to obtain.

We do, however, seem to have been successful at selling the hope that whatever we have—even though is does nothing—may start working at any given time and make its users rich. . . . Now; attacking without building alternatives is never the way to progress, consequently, let us examine how we can modify some macroeconomic models to make them more practical, if at all possible.

The first basic model is called "The Flow of Income" (FI). It is a single equation between two main forces: inflows and outflows to economic activity. Inflows strengthen and quicken the economy, while outflows depress it. The inflows are private investment (I), government expenditure (G), and trade exports (X). The outflows are private savings (S), taxes (T), and trade imports (M). We can express FI as:

(1)          $I(r) + G + X = S(r) + T + M$

Or

(2)          $[I(r) - S(r)] + [G - T] = M - X$

where I, and S are investments and savings, both depending on the real interest rate, r; G, and T are government expenses and revenues; M, and X are imports and exports.

Classical economists, for example, believed that trade balance and balance in the government accounts were two worthwhile goals for political economy since the real rate of interest (r) would keep the private sector in equilibrium by equating investment and saving flows.

FI is a rudimentary but effective teaching device. For example, it does not have a financial side. Still, what it says is very important. During the 1980s, for example, it was common to hear about the "twin deficits"—those of the government and the balance of trade. The point at issue was this: if a country is experiencing healthy growth and strong capital investment, the funds must come from somewhere. At the time, government was in the red and saving flows into the United States were insufficient to finance growth. Therefore, the balance of trade had to experience a deficit to provide the necessary financing.

What we could call the Simple Model of Income Determination (SMID) is associated with the work of John Maynard Keynes and his *General Theory of Employment, Interest, and Money*. Keynes (1883–1946) lived in an age both exciting and challenging because of the world wars and the economic depression of the 1930s. The main features of the SMID are presented in Exhibit 3.1. Consumption depends on current disposable income, investments are exogenous and determined outside the model, and the aggregate supply adjusts passively to whatever demand exists. This is what is called a depression model, in which the government has the power to increase output seemingly at will. For example, if the government were to eliminate taxes (t being the marginal tax rate, make t=0), income (and employment) would double (Y eq = 600) at the cost of a government deficit of 20 units.

**Exhibit 3.1**
**Simple Model of Income Determination (SMID)**

Y demand = C + I + G

Equilibrium: Y demand = C + I + G = Y supply = Y equilibrium

Numerical example:

Income or GNP = Y
Consumption = C = 80 + 0.8 Yd
Disposable Income = Yd = (1-tax rate) Y
Tax rate = t = 0.25
Investment = I = 20
Government Expenditures = G = 20

$$Yeq = \frac{1}{1 - [0.8(1-0.25)]} \, [80+20+20] = 1/4 \, [120] = 300$$

Government Deficit = t Y - G = 0.25 x 300 - 20 = (+)55: a Surplus

---

The SMID has no foreign sector, no supply side, and no monetary sector. These are serious shortcomings because they preempt the existence of inflation, supply side shocks, and monetary policy. The IS-LM model aims at solving some of these deficiencies.

Exhibit 3.2 summarizes a popular version of the IS-LM model, which can be found in most textbooks, and is perhaps the most popular macroeconomic model in business instruction because it is both manageable and allows us to study important issues.

Note how the model stresses major links; the link real-monetary side is the interest rate (i); the link between aggregate demand and aggregate supply is the price level (P). These links, and the overall level of detail in the model, permit some of the most informative assessments of economic policy analysis in the classroom.

**Exhibit 3.2**
**A Simple IS-LM Model**

Real Side:
  Consumption = C = 0.8 Yd = 0.8 (1-t) Y
  Tax rate = t = 0.25
  Investment = I = 900 - 50 i
  Interest Rate = i
  Government expenses = G = 800

Monetary Side:
  Money Demand = Md = 0.25 Y - 62.5 i
  Money Supply = Ms = M/P = 500
  Price Level = P = 1 (scaled to 1 for convenience)

Equilibrium in the real side: The Income-Saving (IS) curve, Yis

  Yis = C + I + G

$$Yis = \frac{1}{1 -[0.8\,(1-0.25)]}\,[\,1700 - 50i\,] = 4250 - 125\,i$$

Equilibrium in the monetary side: The Loans-Money (LM) curve, Ylm

  Md = Ms  -->  500 = 0.25Y - 62.5 i

  Ylm = 2000 + 250 i

Economic Equilibrium: Yis = Ylm

  4250 - 125 i = 2000 + 250 i;  2250 = 375 i  --> i = 6; Y = 3500

Government Deficit = 875 - 800 = 75  (Surplus)

---

An expansive fiscal policy increases government expenditures or decreases taxes. This increases aggregate demand directly through government purchases of goods and services or via an increase in private consumption. But this is simply the initial effect. If the supply side adjusts passively, as in the case of the SMID, all is well. For example, if G increases to 950, the new equilibrium values for income and the interest rate would be Y = 3750, and i = 7. A substantial 7.14% increase in national income (250 over 3500) is acquired at the cost of an additional percentage point in the interest rate and a government deficit of 12.5 units (Deficit = 937.5 - 950 = -12.5).

An expansive monetary policy would increase the money supply. For example, let us make the new money supply, Ms, equal to 593.75. This would yield the following new equilibrium values for income and the interest rate: Y = 3625, and i = 5. Income levels go up, interest rates go down, and even the government shows a surplus (T - G = 906.25 - 800 = 106.25). If this looks too good to be true, it is. We are neglecting something serious: the supply side.

If the economy is at its productive limit (vertical supply curve), an increase in aggregate demand triggered by fiscal or monetary policies will increase prices, depress the real monetary supply (and households' real balances), increase interest rates, and hurt investment. In the worst-case scenario, an increase in government expenses may fully "crowd out" private investment. The ideological side of the issue reappears: what size of government do we want?

In general, we can write, from the equation of the real side of the economy, Y = C + G + I, which implies, differencing, dY = dC + dG + dI. While income grows we may not worry much about the size of the public sector. When income stagnates, dY = 0, increases in government expenses come at the cost of the private sectors (dG = - dC - dI) which, according to historical evidence (for example, former Soviet Union) does not bode well for either the economy or political freedom.

A problem with this version of the IS-LM model is, therefore, its lack of formulations concerning the supply side, the foreign sector, and the labor market. In addition, financial markets are simplified to the extreme because there is only a money market, which prevents studying the term structure of interest rates and cost of capital issues for firms. Term structure models are dynamic in nature, while the IS-LM framework is static. The list of limitations does not end here. For example, two important links—that between investment and the stock of capital, and that between consumption and employment, both via the production function—are absent as well, as Fair has noted (1984, p. 93–94).

The lack of a supply side and the static nature of the model limit its use in studying dynamic problems such as inflation. A major advantage of the model is, however, that it is formulated as a simultaneous equation system:

$$(3) \quad \begin{vmatrix} 1 & -125 \\ 1 & 250 \end{vmatrix} \begin{vmatrix} Y \\ i \end{vmatrix} = \begin{vmatrix} 4{,}250 \\ 2{,}000 \end{vmatrix}$$

This means we can perhaps address some of the deficiencies of the model by adding errors, which may make it more realistic and practical. Before we do so, we must test how far IS-LM models can take us. Under some circumstances, we can make dynamic IS-LM models by linearization around equilibrium values, or via Samuelson's (1983) correspondence principle, which says that the dynamic behavior of a system is presupposed in its static specification.

## C. Looking for a Comprehensive Macroeconomic Model

Our previous analysis suggests there are three major areas in macroeconomic modeling: (1) the interplay between interest rates and income, (2) aggregate demand-supply relationships, and (3) the labor market. The first area informs about the two major causation channels between the economy and its agents: interest rates representing financial conditions and income representing purchasing power. The second area is critical because it represents the link between income growth and employment through the production of goods and services.

We are trying to find a version of the IS-LM model that is acceptable from a theoretical viewpoint; one that allows us to learn from its analytical manipulation and is manageable enough to become practical. This is an ambitious undertaking since we can choose different amounts of variables, some of which could work as endogenous or exogenous depending on our assumptions, and that admit very different functional relationships. The only way to accomplish this task is to keep in mind our objectives, for which we need a set of endogenous variables and a set of exogenous variables to cover income-interest rate, and aggregate demand-supply relationships. This means we are looking for models that include at least income, interest rate, and prices as endogenous variables, and at least policy variables (G, T, money supply) as exogenous ones. Note that policy variables may have their own dynamics and change somewhat independently of policy makers' best intentions.

Exhibits 3.3–3.6 present several formulations that seem amenable to reformulation in approximate form.

Before we make specific comments on each model appearing in the exhibits, we must make some of a general nature.

The models in Exhibits 3.3–3.6 are of the form A x = b, that is, the usual SES form. However, note that functional relationships are written in a generic manner—for example, f(x, y) = z, instead of choosing specific functional forms. To build the matrix structures, we have assumed the model was in equilibrium, and then, we have linearized the model about its equilibrium values. This is how we arrive at A dx = db, which is a simplification of perturbation analysis that assumes dA = 0.

This is important because it makes a critical assumption about the behavior of the system (dA = 0), but also because dynamic analysis is severely hampered by not knowing the functional forms themselves. A well-defined system may have an equilibrium from which we can build its static form via the correspondence principle—Samuelson (1983). However, it is very difficult to make inferences about the system from an arbitrary static form without information about the original functions. Finally, these models have all been built deductively by adding or subtracting hypotheses to the basic supply-demand relationship.

Exhibits 3.3–3.6 present a simple classical model, a simple Keynesian model, a more complex classical model, and a more complex Keynesian model, respectively. They summarize much of the material found in advanced books on macroeconomics and econometrics. Therefore, it will be very difficult for our comments to be fully descriptive. For the purposes of this monograph and this chapter, we must clarify at least the following major issues:

1. Exogeneity. It is important to determine what variables are endogenous and exogenous and which ones are excluded. Excluding a variable (and its functional relationship) such as the labor supply, or the aggregate supply, presumes this variable adjusts passively to the values of other variables. For example, when workers look at the real wage in order to exchange their services for wages (classical case) aggregate supply is determined in the labor market without regard to government policies. When labor supply is excluded (Keynesian case), stabilization policy becomes effective.

2. Specification of functional relationships. The consumption and investment functions in the complex models include more hypotheses about causal relationships than simple models. Aggregate consumption (C ) is made to depend on disposable income ($Y - T - \delta K - ((M+B)/P) \pi$) and be almost as sensitive to financing conditions as investments themselves. The symbol $\pi$ represents inflation. Investments are modeled via Tobin's q, which is a ratio between the present value of investment projects and their market price. It can be expressed as $q = (Fk - ( r + \delta - \pi))/(r - \pi)$, and embodies profit maximization conditions for capital. In this way, aggregate production is linked to aggregate investing: firms will invest if $q > 1$. Both C and I depend on real rates ($r - \pi$) in the classical model.

3. The links between consumption and government's debt (B) monetization are important. Would consumers accept any debt increase, "no matter what"? The link between investments and the behavior of the capital stock is also important and is nowhere to be seen in the simple models. Complex models also split aggregate investment into "net" (ampliation of K) and "maintenance" (replenishment of K at the $\delta$ rate) investment.

4. The classical models exhibit what is called "dichotomy," that is real output is determined in the inputs (K, L) markets independently of aggregate demand. Government can only increase prices by stimulating aggregated demand. Related to the dichotomy is what is called "monetary neutrality," which states, "money is a veil" and cannot influence real processes.

5. Business agents may not care or be able to make inferences about the labor market. However, it is one of the centerpieces of the policy maker. The Keynesian model is a "depression" model, and assumes business expectations are hurt. Consequently, the capital stock is not supposed to increase (dK = 0), which is also assumed in the short term.

6. Nothing is said about the budget process, which often does not seem to be related to purely economic considerations.

7. Prescriptions concerning economic policy effectiveness, crowding out, and price behavior depend on our own assumptions regarding functional forms and response coefficients ($a_{ij}$'s).

The last point is very important. Given the current state of knowledge, most important economic policy assessments depend not on facts but on the perception and implicit or explicit assumptions of the analyst. One of these assumptions is that the system is close to equilibrium.

The models we have presented in the exhibits are still limited, and any other specifications will also be either limited or overburdened by detail. Perhaps the most serious limitation of the complex models is that they do not include financial markets sophisticated enough to allow for a term structure of interest rates, which severely affects modeling the capital budgeting problem by corporations. We have already noted the dynamic limitations of the models we have selected, but it is appropriate to note that assuming dA = 0 also eliminates the private sector response to capricious government policies (Lucas' critique of Keynesian models), which severely limits them for studying expectational issues.

Our lack of knowledge regarding basic assumptions, and the limitations and omission of these models suggests reformulating them in approximate form. As we will see, this reformulation changes our understanding of these models and of their abilities. Before we do so, however, we must examine one area whose discussion has intentionally omitted until now: econometric models. First, because they are supposed to be the practical implementation of theoretical models. Second, because after roughly 70 years of existence they should have prduced a wealth of models and parameter estimates we could use.

### D. The Econometric Connection

Sargent (1987) represents, in our judgment, the state of the art in mathematical macroeconomic modeling. Some of the models in Sargent (1987) cannot be examined in a monograph centered on simultaneous equations systems of the form A x = b, especially those models including rational expectations (see also Begg, 1982). These models require both economic and statistical considerations since they posit rational agents make use of available data in a statistical way, that is, making use of variances and expected values in a way that they do not make systematic forecasting mistakes.

It seems an examination of macromodeling would not be complete without scanning some of the econometric literature for potential tips, hints, and perhaps models we could reformulate in an approximate way. Our search, however, was unsuccessful for the reasons we outline below. We could find no econometric models of complexity similar to that of the complex Keynesian model that we could reformulate in an approximate way.

First, there is the issue of the degree of complexity in econometric models. A few (exact) equations may be insufficient to represent a complex reality, but several hundred equations go beyond what the average firm or institutional investor may want or may be able to handle at the present. For example, Fair (1984) notes that the general trend in econometric modeling seems to favor smaller models than those used by himself and other commercially available models. Some of the later are those of DRI, Chase, Wharton, Brooking, BEA, and so on. Descriptions of some of these are available but some of their content is proprietary.

**Exhibit 3.3**
**A Simple Classical Model**

Formulae

| | |
|---|---|
| $W/P = Fn$ | In the labor market, the labor demand (real wage) equals marginal productivity of labor. |
| $NS = N \, (W/P).$ | The equation for the labor supply is a function of the real wage. |
| $Y = F(N, K)$ | The production function; K is capital and N is labor. |
| $T + S(r) = I\,(r) + G$ | The IS-LM relationship (remember the FI model). |
| $Y = (M/P) \, k$ | A monetarist money relationship. |

Taking differentials

$$d(W/P) = Fnn \; DN + Fnk \; dK$$
$$dN \quad\quad = N' \; d(W/P)$$
$$dY \quad\; = Fn \; dN + Fk \; dK$$
$$dT + S' \; dr = I' \; dr + dG$$
$$dY \; P + dP \; Y = k \; dM$$

Rearranging exogenous and endogenous variables in matrix form, we obtain:

$$
\begin{vmatrix}
1 & -Fnn & 0 & 0 & 0 \\
N' & 1 & 0 & 0 & 0 \\
0 & -Fn & 1 & 0 & 0 \\
0 & 0 & 0 & (S'-I') & 0 \\
0 & 0 & P & 0 & Y
\end{vmatrix}
\begin{vmatrix}
d(W/P) \\
dN \\
dY \\
dr \\
dP
\end{vmatrix}
=
\begin{vmatrix}
Fnk \; dK \\
0 \\
Fk \; dK \\
dG - dT \\
k \; dM
\end{vmatrix}
$$

- Fn, Fnn, Fk, Fnk are the first and second partial derivatives of production to changes in N and K, respectively, and cross derivatives.
- S', I' are the partial derivatives of saving and investment to income.

**Exhibit 3.4**
**A Simple Keynesian Model**

Formulae

$W/P = Fn$                    In the labor market, the labor demand (real wage)
                              equals marginal productivity of labor. There is no
                              labor supply in this model and $dW$ is considered
                              exogenous.
$Y = F(N, K)$                 The production function; $K$ is capital and $N$ is labor.
$Y = C(Y - T) + I(r) + G$     The IS-LM relationship.
$M/P = m(Y, r)$               A Keynesian money supply-money demand
                              relationship.

Taking differentials

$$d(W/P) = dW/P - dP \ (W/P^2) = Fnn \ DN + Fnk \ dK$$
$$dY \quad = Fn \ dN + Fk \ dK$$
$$dT + s' \ dr = I' \ dr + dG$$
$$dM \ 1/P + (M/P^2) \ dP = my \ dY + mr \ dr$$

Rearranging exogenous and endogenous variables in matrix form, we obtain:

$$
\begin{vmatrix}
Fnn & 0 & 0 & W/P^2 \\
-Fn & 1 & 0 & 0 \\
0 & (1 - C') & -I' & 0 \\
0 & my & mr & M/P^2
\end{vmatrix}
\begin{vmatrix}
dN \\
dY \\
dr \\
dP
\end{vmatrix}
=
\begin{vmatrix}
(dW/P) - Fnk \ dK \ (*) \\
Fk \ dK \\
dG - C' \ dT \\
dM \ 1/P
\end{vmatrix}
$$

$C'$, $I'$, are the partial derivatives of consumption and investment to income.

(*)    This equation can be expressed in different ways. We start with
       $dW \ 1/P - w/P^2 \ dP = Fnn \ dN + Fnk \ dK$, from where we obtain
       $Fnn \ dN - w/P^2 \ dP = -dW \ 1/P + Fnk \ dK$. Noting that $W/P = Fn$
       and changing signs, we arrive to $Fnn \ dN + Fn/P \ dP = +dW \ 1/P - Fnk$
       $dK$ ; or $Fnn/Fn \ dP + 1/P \ dP = dW/W - Fnk/Fn \ dK$, as in the
       corresponding equation of Exhibit 3.6.

**Exhibit 3.5**
**A More Complex Classical Model**

---

Formulae

a)  $W/P = Fn$
b)  $N = N(W/P)$
c)  $Y = F(N, K)$
d)  $C = C(Y - T - \delta K - ((M+B)/P) \pi + (q(K, N, r - \pi, \delta) - 1) I, r - \pi)$
e)  $I = I(q(K, N, r - \pi, \delta) - 1)$
f)  $Y = C + I + G + \delta K$
g)  $M/P = m(Y, r)$

Matrix expression after taking derivatives and noting DM = DB (open market operations constraint)

$$
\begin{vmatrix}
1 & -Fnn & 0 & 0 & 0 & 0 & 0 \\
-N' & 1 & 0 & 0 & 0 & 0 & 0 \\
0 & -Fn & 1 & 0 & 0 & 0 & 0 \\
0 & -C1Iqn & -C1 & 1 & -C1(q-1) & a_{4,6} & a_{4,7} \\
0 & -I'qn & 0 & 0 & 1 & -I'q_{r-\pi} & 0 \\
0 & 0 & 1 & -1 & -1 & 0 & 0 \\
0 & 0 & my & 0 & 0 & mr & M/P^2
\end{vmatrix}
\begin{vmatrix}
d(W/P) \\
dN \\
dY \\
dC \\
dI \\
dr \\
dp
\end{vmatrix}
$$

where $a_{4,6} = -(C1Iq_{r-\pi} + C2)$, $a_{4,7} = -C1 \pi ((M+B)/P^2)$,  and the vector of exogenous factors is

$$
= \begin{vmatrix}
Fnk\ dK \\
0 \\
Fk\ dK \\
[-C1\ dT - C1\ \delta\ dK - C1\ ((M+B)/P)\ d\pi + I\ q_k\ dK - (C1\ I\ q_{r-\pi} + C2)\ d\pi] \\
I'\ qk\ dK - I'\ q_{r-\pi}\ d\pi \\
dG + \delta\ dK \\
dM/P
\end{vmatrix}
$$

---

*Source:* Sargent (1987, Chapter I). © Academic Press. Used with permission.

**Exhibit 3.6**
**A More Complex Keynesian Model**

Formulae

a)  $Y = F(N, K)$
b)  $W/P = Fn$
c)  $C = C(Y - T - \delta K - ((M+B)/P) \pi, r - \pi)$
d)  $I = I( q(K, N, r - \pi, \delta) - 1)$
e)  $Y = C + I + G + \delta K$
f)  $M/P = m(Y, r)$

The matrix expression after taking derivatives and noting $d\delta = 0$, and DM = DB (open market operations constraint) is

$$
\begin{vmatrix}
1 & -Fn & 0 & 0 & 0 & 0 \\
0 & Fnn/Fn & 0 & 0 & 0 & 1/P \\
-C1 & 0 & 1 & 0 & -C2 & a_{4,7} \\
0 & -I'qn & 0 & 1 & -I'q_{r-\pi} & 0 \\
1 & 0 & -1 & -1 & 0 & 0 \\
my & 0 & 0 & 0 & mr & M/P^2
\end{vmatrix}
\begin{vmatrix}
dY \\
dN \\
dC \\
dI \\
dr \\
dP
\end{vmatrix}
$$

where $a_{4,7} = -C1 \, \pi \, ((M+B)/P^2)$, and the vector of exogenous factors is

$$
=
\begin{vmatrix}
Fk \, dK \\
dW/W - (Fnk/Fn) \, dK \\
-C1 \, dT - (C2 + C1 \, ((M+B)/P)) \, d\pi \\
I' \, qk \, dK - I' \, q_{r-\pi} \, d\pi \\
dG + \delta \, dK \\
dM/P
\end{vmatrix}
$$

*Source*: Modified from Sargent (1987, Chapter II), who also assumes dK = 0. © Academic Press. Used with permission.

Those descriptors may be found in From and Klein (1975), Klein and Young (1980), and Spivey (1979). Fair's (1984, p. 1) observation that there is a move away from large-scale macromodels seems to echo Morishima's gracious way of comparing their model to larger ones: "(C)ompared with the monumental Brookings model of Duesenberry, Fromm, Klein, Kuhn and others, the present model may be a midget, but the world was not made only for giants" (Morishima et al., 1972, p. 2).

Second, trying to get a good fit often obscures the very things we would like to learn about. For example, when different error specifications or different lag structures are used.

Third, the econometric specification can be written as $Y \Gamma i + X Bi + ei = 0$, see Judge et al. (1985), or Maddala (1977). We could also write

(4)     $A Yt + B Y t\text{-}i + C Xt + D Xt\text{-}i + Vt = 0$

which differs from the SES specification $A x = b$ in two important respects. First, in the potential presence of lagged values of endogenous and exogenous variables. The presence of such terms was justified on the basis of (1) simply trying to get a better fit, or more recently, (2) because they may represent expectational hypotheses, or (3) they appear naturally as we employ certain types of regression techniques. It is fair to say that mapping such terms to the theoretical economic hypotheses of the previous sections is very difficult. Second, the presence of the error term is misleading from a theoretical (or analytical) viewpoint. In econometrics that error is supposed to be well behaved, otherwise it may "take over" the model itself. However, if the error is simply a transparent veil over exact processes, it will not add anything to the theoretical content of the model.

Fourth, in other occasions econometric specifications are formulated in conjunction with optimization techniques such as dynamic control, see Chow and Corsi (1982). In this case, the theoretical content of the model gives in to the study of the technique being presented.

Fifth, time-series multi-equation macromodels play a major role in econometric research. These are for the most part noncausal models and therefore we have omitted them from our search for practical, reasonably rich, macroeconomic models. The interested reader is referred to Judge et al. (1985), which we consider the best survey on the state of art in econometric techniques, and to the excellent texts by Maddala (1977), and Johnston (1984). Begg (1982) is a very informative and clear text that highlights the special econometric problems arising from alternative expectational hypotheses. Vector autoregressions (VAR), including ARCH and GARCH specifications, are also a very effective tool to forecast single variables, which are used in many areas of business, academic, and public policy research. We did not include them in our survey because they are mostly uniequational, and do not include causality relationships.

We also reviewed some monographs on the borderlines between macroeconomic theory, economic policy, and econometrics, but these texts rarely use a moderately complete model (such as the complex models presented earlier) to make their points. For example, Barro (1990) covers many points concerning macroeconomic policy making his point with either purely theoretical models of a couple of dynamic equations and no data, or by partial models associating a few variables or concentrating on a very specific segment of the money market. We cannot imagine how anyone could study general macroeconomic policies under such conditions. UN-ECEE (1967) and Morishima et al. (1972) are better examples of practical econometric model building. Unfortunately, the models employed in these sources are not amenable to being reformulated in approximate form because of the first three reasons given earlier.

Surprising as it may seem, the business person interested in adapting some a priori sound model for practical use will find there are no econometric implementations of models such as those presented by Sargent (1987). This fact invites some further reflections:

First. How come? This is the most obvious question and one possible answer, which has many implications, is that the current level of knowledge about macroeconomic relationships is simply insufficient to implement such apparently theoretical sound models. Sargent himself does not provide any numerical example for his macromodels in his otherwise excellent text. In using plausible values—as noted in the following section—it is apparent that the models are very fragile and have a tendency to (1) not have room for all the effects they seem to encompass in theory, (2) provide results that are extremely sensitive to very small changes in the inputs, and (3) are threatened by ill conditioning, as the previous point suggested.

Second. How can one build econometric models with no theoretical support? And, if one goes ahead and obtains a good fit with a variety of lagged endogenous and exogenous variables here and there, how can one assure there is no overfitting when the economic theoretical support is removed? Current economic research is facing the first question, and the second question is addressed by research in econometrics and statistics. The reader may judge for herself or himself how well this research is answering these questions.

One general implication of these observations is that economic policy making is mostly based on ideology in spite of positivistic pretensions of the economic profession. Another implication is that there is economic knowledge that agents may be using and theoretical macromodels cannot handle.

Do our findings in this first segment of this chapter indicate that macroeconomic modeling is not ready to help in practical situations? In a way it does indicate so, but this is highly counterintuitive for us. There must be ways in which business-people and investors can build manageable, nontrivial models that can be used to reasonably justify their decisions and build expectations regarding the economic environment. The way to do so is to focus on developing models that are practical to users, instead of focusing on policy making and gigantic, resource and computational-intensive models. This practical way also entails including errors in our theoretical formulations. This monograph tries to progress along the road to practical applications, but it is not the only one. Klein (1950) presents one such model. The next section follows, Naylor's (1971) analysis of Klein's six-equation model.

### E. Beyond Quantitative Models

In addition to the models referred to in Sargent (1987), others can be found in Begg (1982), and in Cambell, Lo and MacKinlay (1997), who provide a thorough overview of the state of the art in financial econometric modeling, which believe it or not, does not include macroeconomic modeling.

One cannot help being very surprised at the state of conventional micro- and macroeconomic modeling (econometric or not), which seems filled with puffery and ornamentation, yet has little regard for the practical needs of its users. Fischer Black (1982), one of the most respected researchers in finance, expressed his frustration with econometric models in an article published in the Financial Analyst Journal, which has a wide audience including security analysts. His conclusion is clear and clearly stated: "It is doubtful that traditional econometric methods will survive" (1982, p. 29).

Black's analysis starts by noting that it is usually not possible to estimate even supply and demand relationships. The major problem is that of ascertaining causality, without which it is impossible to build a structure. Without the grounding of causality, the relationships established are not sound, the analyst confuses correlation with causality, and the resulting model is useless. Furthermore, these models are kept alive on the basis that "(I)t is easier to use correlations for forecasting than it is to interpret them. If we are using them for forecasting, we do not have to understand them" (1982, p. 34).

At this point the question is this: Does anyone question the fundamentals any longer? If we come up empty-handed from fundamental macromodels that should be at the core of sophisticated models, what do we expect to gain? What do we actually gain? Is performance analysis ever applied to our methodologies and research strategies? These questions motivated this monograph indirectly. We believe that (1) we have not learned enough from fundamentals, (2) that the failure of current econometric modeling stems from a senseless search for computational wizardry and disregard for alternatives to standard probability theory. We submit the lack of practical results and minor usefulness of (published) macroeconometric models as evidence.

It also happens that economic knowledge goes well beyond quantitative models, see Fiedler (1984), Kuttner (1985), and Thurow (1984) for examples. Exhibit 3.7 summarizes macroeconomic representations beyond those purely quantitative, which appear on the right-hand side of the exhibit. When we migrate from numbers to qualitative relationships, we observe that we enter the most populated grounds of economic analysis, at least by practitioners, business-people, households, and politicians. Even professional portfolio managers favor qualitative analysis based on principles rather than quantitative models, see Madrick (1988).

It seems investors (firm planners, financial intermediaries, households, and so on) may use information contained in these representations to make spending, investing, and resource allocation decisions. The reader of these lines, not to go any further, may or may not be a doctorate in economics or business, yet, he or she makes delicate decisions in an increasingly complex environment. These complex decisions are, to wit, education, firm to work for, home to purchase, mortgage to sign, geographical location, children to raise, and so on. There is no way that each of us, as individuals, could have a fulfilling life in the complex economic environment if we were unable to distill some nectar of economic wisdom when we need it. It is obvious that most of us make a very fine and sophisticated use of economic knowledge. We, as scholars, know very little about individual representations of economic knowledge or the cognitive properties of economic decision-making.

**Exhibit 3.7**
**Representations of Economic Knowledge**

| Key relationships<--> Discursive models <--> Full models <--> Partial models |
| --- |

| Business cycle | Classical model | IS-LM | Money market |
| --- | --- | --- | --- |
| | Keynesian | | Credit conditions |
| | Neoclassical | | Supply-demand |
| | Neokeynesian | | External sector |
| | New-classical | | |

Building causal quantitative models is one of the ways to strengthen individual decision-making that this monograph is studying on approximate equations. If this is so, it seems that basic IS-LM models are good starting points and adding flexibility to the fundamental IS-LM model may improve them. This is the objective of the following section.

## 2. TOWARD PRACTICAL MACROECONOMIC MODELS

This second part of the chapter will walk the road to building practical models. First, it will refocus on the problem it is trying to tackle. Second, it will outline the theoretical implications of reformulating models as approximations. Third, it will examine approximate formulations of Sargent's (1987) (complex) Keynesian model and another interesting model provided by Klein (1950) as it appears in Naylor (1971).

There are two ways to motivate an interval version $AI \ xI = bI$ of the basic model $Ax = b$: (1) build the interval version from scratch with the help of interval arithmetic, and (2) modify $Ax = b$ to accommodate intervals. We favor the second option for reasons such as the availability of a simple method to solve these equations, and because doing so does not require special functional spaces of numerical fields. Adding errors to the exact models is equivalent to relaxing the phenomenal accuracy requirements of exact models.

## A. The Problems and the Agents

This chapter focuses on the problem of dealing with, managing, handling, or surviving the environment. This is the oldest problem known to us as human beings. Contesting or enjoying our many environments (economic, biological, human, and so on) is part of living.

The environment, defined as "everything that is not the individual," is a complex of issues that overwhelms our intelligibility and tools. Consequently, we parcel the environment into areas that are most directly linked to our immediate subproblems. This case is a study of economic planning problems. Therefore, it concentrates on the part of the environment we associate with economic issues.

Income and financing are the two most commonly used channels of economic causation. Income is ultimately related to wealth: it is the flow that fills the container of wealth that, in turn, is the stock of accumulated resources. Our economic well-being is related to the level of the wealth container as well as the flow of income. Note that along with the inflow of income, the wealth container will also experience leaks representing allocations to current production (consumption, working capital management) and investment.

Economic agents are first of all planners, which also highlights the role of the external environment in decision-making. For example, there is evidence that the aggregate economy may explain up to 50% of variation in the firm's total income, see Elton and Gruber (1995, pp. 484–485). Assessments of aggregate economic performance seem very important in earnings predictions. Furthermore, in a study of individual (as opposed to institutional) investing, Jacob (1974) noted that there was a need for low-cost forecasting models that individuals could use to support their decision-making. As we have seen, there is a great deal to be learned about economics before developing such a practical model, which ideally should be adaptable to the individual's cognitive capacity and background (customizable). Also remember Schopenhauer (1958, p. 56) observed that practical means intuitive. In this sense, this study is making the rational models of Exhibits 3.3–3.6 intuitive and, therefore, practical.

Even on a priori terms, the very fact that we cannot handle detailed models with tens of variables and relationships suggests formulating in an approximate way whatever simpler construct we may have. Our goal may need to be modest, given the formidable uncertainties we face in macroeconomic modeling.

## B. AES Specifications: Implications

Our choice of model is determined by what we most want to learn about. In this case, it is the interplay between financing and income conditions. This can be done with the help of the IS-LM model presented in Exhibit 3.2, and is informative by itself. The problem is that the critical element in the forecast, the supply side, is left out. But including aggregate supply brings in considerations regarding labor and capital (K) markets. Another pertinent consideration is that we know our (exact) static model may double up as a dynamic form under certain conditions. Then how many elements can we include? The general answer to this question is always the same and is referred to as "Ockham's razor": use as few elements as possible. Therefore, we will use those variables that matter for the individual, namely income, interest rates, prices, inflation, and employment.

Adding errors has very important implications for macroeconomic modeling. Take first, for example, considerations regarding the external environment. Once we add errors to the vector $b = \{ b1, b2, b3, \ldots, bk \}$ we do not know exactly which variables are changing, or which changes are compensating for what. Furthermore, changes in economic policy (changes of G, T, or money supply) will coexist with other changes in the environment.

Adding errors also has critical effects on the internal world of the model (matrix A). First of all, the models no longer necessarily have a dichotomy between real output and aggregate demand. For example, the dichotomy in the classical models of Exhibits 3.3 and 3.5 is broken when zeroes are replaced with non-zero values in the interval $[ 0 \pm \varepsilon ]$. In the Keynesian model, the corresponding link is $a27 = 1/P$, which can also be zero depending on the size of the error. Adding an error to the coefficients of the original A matrix also eliminates the "splitting of hairs" concerning the sensitivity of the investment and consumption function to changes in interest rates, or the reactions of money demand to changes in government policies, sensitivity of the supply curve, or existence of a natural rate of unemployment.

Adding errors exposes the relativity of our assumption, models, and inferences. But this relativity is productive and forward looking, telling us the limits of our theories, tools, and data. If manageable models are those of only a few equations, then we are likely to miss variables in our specifications. For the same reason, we wouldn't know which or how many exogenous variables were changing at a given time. We do not think we have the theoretical distinguishability to know with precision the values of the coefficient matrix A either.

## C. Two Examples of Approximate Macromodels

In the previous chapter we quoted Bertrand Russell saying that "(E)verything is vague to a degree you do not realize till you have tried to make it precise" (1985, p. 38). Well, this is certainly an understatement concerning macroeconomics.

We selected the complex Keynesian model (Exhibit 3.6), and proceeded to study its numerical properties. That is, we assigned plausible values to each parameter in the model. We found this theoretical specification to be very fragile in numerical and theoretical terms. This fragility shows in three ways: (1) the model is very sensitive to changes in plausible values for coefficients, (2) coefficient matrix A is sensitive to ill conditioning, and (3) it is very difficult to obtain a complete range of even textbook effects without making arbitrary changes in the coefficients.  Textbook effects, such as those following monetary or fiscal policies, can be obtained by strategically placing zeroes, but the more zeros, the more sensitive the matrix becomes to ill conditioning.  We already mentioned our perplexity at finding no numerical examples in Sargent's text.

Whatever the case, we must also note that a given model would be even more fragile when econometric estimates are used, because the "signal" they carry is weaker than that of theoretical plausible values.  The first example will add errors to "plausible values", and the second example will use econometric estimates.

The subsections that follow will present each of the models and the detail concerning their formulation in approximate form.  The next section will contain general observations about approximate results shared by both models.

### 1. Sargent's Complex Keynesian Model

Exhibit 3.8 and 3.9 presents a numerical example of Sargent's complex Keynesian model from Exhibit 3.6.

The values used in our numerical illustration are plausible and further illustrate some important characteristics of macromodeling.  For example, the positive values for marginal productivity of labor and capital indicate aggregate production will increase when any one of these inputs increases.  The second derivatives however, tell us that output will increase at a decreasing rate.  "P" and "W," which are the level of prices and salaries, respectively, have been indexed to 1.23 and 1, respectively.  The marginal propensity to invest due to changes in real interest rates is written in the following way: $I'qr\text{-}\pi = \text{-}$ sensit $*$ (q $-$ 1), where q is defined as we noted in the description of the model and sensit is a coefficient of impact of changes in q on the marginal propensity to invest.

Study of our numerical example shows that the complex Keynesian model incorporates a remarkable richness of economic effects.  For example, the links between (1) wages, labor supply, and consumption, and (2) aggregate investment, productivity of capital, and interest rates, are both very well laid out and effectively integrate supply- and demand-side issues at what could be called the root level of the model.  The model appears to be a solid educational tool as well.  The model also exhibits internal consistency in the sense that changes in some variables propagate as the theory underlying the model predicts.

The matrix of coefficients has a determinant distinct from zero—that is, it is invertible—and the slopes of the IS-LM are negative and positive.  This indicates that they represent a stable system.  The maximum error that can be added/subtracted from the coefficients before the matrix becomes ill conditioned is 0.021646.

The last block in Exhibit 3.8 presents the response of endogenous variables to changes in exogenous variables, some of which implement economic policies.  Each column represents a change in a single exogenous variable, that is "keeping all else equal" (ceteris paribus).  We can briefly comment on the following effects:

- A 1 percent increase in the wage index (an external supply-side shock) depresses aggregate income, employment, consumption, and investment; it also increases prices and interest rates.
- An increase in the stock of capital has the contrary effect to a rise in aggregate wages, with exception of decreasing aggregate investment, perhaps due to a relative exhaustion of investing opportunities.
- A tax increase depresses aggregate consumption, which lowers income and interest rates. This, in turn, stimulates investment and income to partially offset the consumption shock.
- The consumption function is very richly linked in this model, as we can see by the central role played by consumption to changes in the policy variables. For example, note how expansionary effects of monetary policy are ultimately felt through a wealth effect.
- Fiscal policy is rather effective for income expansion purposes, which is consistent with the makings of a Keynesian model. The "crowding out" phenomenon is not strong enough to offset the increase in government expenses. The model also exhibits a positive "balanced budget" multiplier—see the last column.

In general, the model is very sensitive to changes in some variables, and small changes in a few parameters can produce "classical effects." Small changes in parameters can also bring the model dangerously close to ill conditioning. The plausible values we originally input were only slightly calibrated to obtain textbook effects in economic policies without breaking down the model (causing ill conditioning). We must note, however, that there is little room for manipulation of inputs.

Exhibit 3.9 presents the solutions for an approximate version of the model in Exhibit 3.8. This exhibit also highlights the limitations of the technique of linearization, which has its pluses and minuses. For example, it is good for theoretical analysis of qualitative algebraic values, but less useful to detect realistic base values.

There is another clarification we must make concerning the way we obtain solutions for approximate equations in the case of linearizations. When we add errors to the matrix A and the vector b, we may think we should refer to equation (8) of Chapter 2, which concerned perturbation theory, to build our dx estimates $(\Delta x = (A+\Delta A)^{-1} \Delta b - (A+\Delta A)^{-1} \Delta A \ x^*)$. Our examples will not do so because there are no base values in the complex Keynesian case and, more importantly because it is not known whether the size of errors are compatible with perturbation theory. This case illustrates why approximate models would be more properly built from scratch using interval mathematics methods. The problem is that this is a colossal undertaking and we still need some approximations while it is under development.

**Exhibit 3.8**
**Sargent's Complex Keynesian Model**

Coefficients and initial data:

Fnk = -0.04; Fn = 0.559335; Fk = 0.221684; Fnn = -0.00543; P = 1.233933

W = 1; C1 = 0.8; C2 = -0.04879; M = 7.488155; sensit = 2.775912

I'qn = -0.11626; I'qk = -0.1; my = 0.8; mr = -0.8; r = 0.08 ; delta = 0.05

$\pi$ = 0; q = 1.146053; I'qr-$\pi$ = -0.40543; M + B   = 8.23697

dW = dK = dT = d$\pi$ = dG = dM = 0

|              | Matrix A |       |       |          |          | Vector b |
|-------|-------|-------|-------|-------|-------|-------|
| 1     | -0.55934 | 0     | 0     | 0        | 0        | 0        |
| 0     | 0.009701 | 0     | 0     | 0        | 0.810417 | 0        |
| -0.8  | 0        | 1     | 0     | 0.04879  | 0        | 0        |
| 0     | 0.116257 | 0     | 1     | 0.40543  | 0        | 0        |
| 1     | 0        | -1    | -1    | 0        | 0        | 0        |
| 0.8   | 0        | 0     | 0     | -0.8     | 4.918035 | 0        |

Slope IS = -0.89791; Slope LM = 1.200322

Det. A = -0.29095; Max error 0.021646

|     | DW<br>= 0.01 | dK<br>= 0.10 | dT<br>= 0.10 | d$\pi$<br>= 0.01 | dG<br>= 0.10 | dM<br>= 0.10 | dG = dT<br>= 0.10 |
|-----|----------|----------|----------|----------|----------|----------|----------|
| dY  | -0.04295 | 0.028572 | -0.09971 | -0.0609  | 0.12464  | 0.057351 | 0.024928 |
| dN  | -0.07678 | 0.011448 | -0.17827 | -0.10888 | 0.222837 | 0.102535 | 0.044567 |
| dC  | -0.03624 | 0.024151 | -0.15554 | -0.09906 | 0.094431 | 0.048394 | -0.06111 |
| dI  | -0.00671 | -0.00058 | 0.055833 | 0.038155 | -0.06979 | 0.008958 | -0.01396 |
| dr  | 0.038561 | -0.02652 | -0.08659 | -0.05289 | 0.108242 | -0.0515  | 0.021648 |
| dP  | 0.013258 | -0.00896 | 0.002134 | 0.001303 | -0.00267 | -0.00123 | -0.00053 |

**Exhibit 3.9**
**Sargent's Complex Keynesian Model: Approximate Equations Solutions**

|     | Error <br> IC factor <br> Max | 0.01 <br> 0.461989 <br> Min | Error <br> IC factor <br> Max | 0.02 <br> 0.923979 <br> Min |
| --- | --- | --- | --- | --- |
| dY | 0.135602 | -0.13561 | 0.653277 | -0.65328 |
| dN | 0.270785 | -0.27079 | 1.304445 | -1.30445 |
| dC | 0.125525 | -0.12553 | 0.604689 | -0.60469 |
| dI | 0.031916 | -0.03192 | 0.091669 | -0.09167 |
| dr | 0.095811 | -0.09581 | 0.237466 | -0.23747 |
| dP | 0.022797 | -0.0228 | 0.10982 | -0.10982 |

Configuration of max dy, dN, dC, and min dP
Binaries: (A, b)

$$
\begin{vmatrix}
1 & 1 & 1 & 0 & 0 & 0 & \vdots & 0 \\
1 & 1 & 1 & 0 & 0 & 0 & \vdots & 0 \\
0 & 0 & 0 & 1 & 1 & 1 & \vdots & 1 \\
0 & 0 & 0 & 1 & 1 & 1 & \vdots & 1 \\
0 & 0 & 0 & 1 & 1 & 1 & \vdots & 1 \\
0 & 0 & 0 & 1 & 1 & 1 & \vdots & 1
\end{vmatrix}
$$

Configuration of min dy, dN, dC, and max dP
Binaries: (A, b)

$$
\begin{vmatrix}
0 & 0 & 0 & 0 & 1 & 1 & \vdots & 1 \\
1 & 1 & 1 & 1 & 0 & 0 & \vdots & 0 \\
1 & 1 & 1 & 1 & 0 & 0 & \vdots & 0 \\
0 & 0 & 0 & 0 & 1 & 1 & \vdots & 1 \\
1 & 1 & 1 & 1 & 0 & 0 & \vdots & 0 \\
0 & 0 & 0 & 0 & 1 & 1 & \vdots & 1
\end{vmatrix}
$$

## 2. Klein's (1953) Model

Exhibit 3.10 presents Klein's (1953) model as it appears in Naylor (1971). We found Klein's model interesting for several reasons. First, it seems consistent with a "going back to the foundations." Second, it seems to be built with practical purposes in mind in terms of its manageable number of equations and data requirements. Third, it further illustrates a point of practical importance: calculation of model parameters.

As we know, parameters can be calculated in at least three ways: (1) econometrically as Klein does, (2) by simulation methods, and (3) analytically. Naylor (1971, pp. 4, 143, 181, and the entire twelfth chapter) finds this choice critical. The best estimates, in his opinion, are analytical, or exact, solutions obtained from theoretical relationships—the plausible-plus-calibrated estimates from the previous example. Next best are econometric optimization techniques. This leaves simulations as a "last resort" option.

The problem with analytical solutions is that they may represent all cases, in theory, but misrepresent some of the most important ones, since in economics and finance the average case is not always the most important. The problem with econometric techniques is that we may be obtaining good fits by altering the model in a way that does not make theoretical sense, even with mis-specified models, and the parameters we obtain might be highly unstable, as we noted earlier. Simulation amounts to a calibration of parameter values and errors. This calibration makes the final model usable. It must be inferior to the other two when informational content of the model is good and the model is linear, but it may be advantageous when such information is deficient. Klein's model will be used by keeping the regression or probabilistic errors, but while omitting residuals and working with exact form equations, to which mathematical errors will be added. This amounts to working with the model in its expected form. (Let $yt = a + b\ xt + ut$ be the typical regression equation where $ut$ is the usual white noise error, which has an expected value of zero. Then the expected value of $yt$ will be $yt = \alpha + \beta\ xt$, or $yt+1 = \alpha + \beta\ xt+1$, where $\alpha$ and $\beta$ are the usual regression coefficients.)

The fourth interesting aspect of Klein's model is that the way the model is used is instructive for our purpose of developing practical models. Once the parameters have been estimated, the model is used to simulate values for income when other values are changing. As we saw earlier in Chapter 2, given $A\ x = b$, we can calculate $x = A\text{-}1\ b$, and posit $x + dx = A\text{-}1\ (b + db)$ if we assume that the coefficient matrix A does not change. Different runs of values for exogenous variables are used to estimate change for the endogenous variables, which is calculated by, for example, averaging trial values to a single-point estimate.

In dealing with approximate equations, we obtain a range related to the error-carrying capacity of the structure we are handling. We cannot determine how useful simulations are if we have not first studied the reliability of the measurement tool—the model itself. In any case, simulation techniques are plagued by problems such as determination of sample size, structural change, large variance in estimates, multiple response to changes in exogenous variables, interdependence, nonlinearity, and other model-specification errors. All of these problems strengthen the case for using approximate equations models in place of, or in addition to, simulation.

Exhibits 3.10 and 3.11 contain Klein's model, its original algebraic equations, and a linearization around equilibrium values. From the viewpoint of modeling, existence of ones and zeroes can be a harbinger of disappointment. Ones may express identities and, as such, are uninteresting because they are most binding. Zeroes may indicate lack of links, which may weaken the informational fabric as if they were holes. Klein's model has 6 ones and 20 zeroes. The complex Keynesian model also has 6 ones and 20 zeroes—excluding those in the vector b. Therefore, both models seem to be of comparable quality: the complex Keynesian stronger on the side of theory, Klein's on that of data.

The determinant of the coefficient matrix is different from zero, $\det(A) = 0.616529$, and the maximum error that can be added/subtracted from the coefficients before the matrix becomes ill conditioned is 0.047092.


## 3. Analysis of Approximate Equations Solutions

We have made a long journey into macroeconomic modeling and mathematics in order to formulate models in approximate form. We must refresh the answers to two questions: What do these errors mean? How do they improve exact models?

These errors are an indicator of the error-carrying capacity of the mathematical container that holds information. The strength of this container is based on our knowledge of the problem, which manifests itself in the values of the coefficients in the matrix A—the structure of the problem itself. When we do not know enough, the model we build is weak and the objects in which it is housed are also weak and have a tendency to crack and break down (critical ill conditioning). Presumably, the best models would be those far away from being ill conditioned, and those in which our knowledge of the problem is so good that mismeasurement and imprecision have little effect on the applicability and usefulness of the model. For example, it is well-known that a system is ill conditioned when small changes in the inputs produce very large changes in the output of the model that are unjustified from a theoretical standpoint.

With respect to the second question, it is obvious our models are imperfect for myriad of reasons, many of which have already been listed in the previous two chapters. This short, but critical, survey of macroeconomic modeling should serve to stress how much we need to learn in macroeconomic model-building. We should also remember that 99 percent of economic and financial model-building is done for forecasting purposes, and that our ignorance increases when we deal with the future. Approximate equations link the presence of errors to obtaining ranges of uncertainty for the variables involved: this is critical if the models are to have any expectational value at all.

**Exhibit 3.10**
**Klein's (1953) Model**

Goal: Build a model to examine the effects of one or more government fiscal policies on economic behavior in the United States.

Policy variables
$W2t$     = governmental wage bill in period t
$Gt$       = governmental demand in period t
$TXt$     = business taxes in period t

Output variables
$Ct$       = consumption in period t
$W1t$     = private wage bill in period t
$\Pi t$       = non-wage income (profits) in period t
$It$        = net investment in period t
$Kt$       = capital stock at end of period t
$Yt$       = national income in period t

Behavioral equations in exact (or expected) form
$Ct$   = $a1 + a2 (W1t + W2t) + a3 \Pi t + a4 \Pi t\text{-}1$ = consumption function
$It$   = $b1 + b2 \Pi t + b2 \Pi t\text{-}1 + b4 Kt\text{-}1$ = investment function
$W1t$  = $c1 + c2 (Yt + TXt - W2t) + c3 (Yt\text{-}1 + TXt\text{-}1 - W2t\text{-}1) + c4 T =$
        demand for labor function

Identities
$Yt$    = $Ct + It + Gt - TXt$        = national income equation
$\Pi T$    = $Yt - (W1t + W2t)$        = profit equation
$Kt$    = $K t\text{-}1 + It$            = Capital stock equation

Parameter estimation (via three-stage least squares)
$Ct$   = $16.44 + 0.7901 (W1t + W2t) + 0.1249 \Pi t + 0.1631 \Pi t\text{-}1 + error1$
$It$   = $18.18 - 0.01308 \Pi t + 0.7557 \Pi t\text{-}1 - .1948 Kt\text{-}1 + error2$
$W1t$  = $15.08 + 0.4005 (Yt + TXt - W2t) - 0.1813 (Yt\text{-}1 + TXt\text{-}1 - W2t\text{-}1)$
        $- .1497 T + error3$

---

*Source*: Computer Simulation Experiments with Models of Economic Systems, T. Naylor. Copyright © 1971, John Wiley & Sons, Inc. Reprinted by permission of John Wiley and Sons, Inc.

Exhibits 3.9 and 3.12 deliver the two macroeconomic models in approximate form. We will limit our comments to the most salient features of reformulating the models in approximate form, from which we highlight the following.

**Exhibit 3.11**
**Klein's (1953) Model: Approximate Form**

Exact formulation (without residuals) of linearization around equilibrium values.

$$
A = \begin{vmatrix}
1 & 0 & -.7901 & 0 & -0.1249 & 0 \\
0 & 1 & 0 & 0 & 0.01308 & 0 \\
0 & 0 & 1 & -0.4005 & 0 & 0 \\
-1 & -1 & 0 & 1 & 0 & 0 \\
0 & 0 & 1 & -1 & 1 & 0 \\
0 & -1 & 0 & 0 & 0 & 1(*)
\end{vmatrix}
\quad
x = \begin{vmatrix}
dCt \\
dIt \\
dW1t \\
dYt \\
d\Pi t \\
dKt
\end{vmatrix}
$$

$$
= \begin{vmatrix}
0.7901\ dW2t + 0.1631 d\Pi t \\
0.755\ d\Pi t\text{-}1 - 0.1948\ dK\ t\text{-}1 \\
0.4005\ (dTXt - dW2t) + 0.1813\ (dYt\text{-}1 + dTXt\text{-}1 - dW2t\text{-}1) + 0.1497 \\
dT \\
dGt - dTXt \\
dW2t \\
dKt\text{-}1
\end{vmatrix}
$$

b

(*) This coefficient erroneously appears as a zero in Naylor (1971, p. 131)

*Source*: Computer Simulation Experiments with Models of Economic Systems, T. Naylor. Copyright © 1971, John Wiley & Sons, Inc. Reprinted by permission of John Wiley and Sons, Inc.

Regarding the mathematical side, the size of the ranges of uncertainty is, as should be expected, proportional to the error being considered. Second, in the two models studied, the linearization technique provides solutions that are symmetrical to the centered solution.

Regarding economic implications, what the ranges of uncertainty indicate are possible values for the variables of interest that are consistent with the mathematical objects in which the analysis is carried out. That is, in the complex-Keynesian case, if we think the true value for each coefficient may lie within the interval built by adding and subtracting 0.01 from the coefficients we actually use, then solution values for the variables could be in the spread established by the uncertainty ranges given in the table. For example, with an error of { ±0.01} aggregated income (dY) resides somewhere in the interval [+0.135602,–0.135602].

**Exhibit 3.12**
**Klein's (1953) Model: Approximate Equations Solutions**

| | | | | | | | |
|---|---|---|---|---|---|---|---|
| 1 | 0 | -0.7901 | 0 | -0.1249 | 0 | | 0 |
| 0 | 1 | 0 | 0 | 0.01308 | 0 | | 0 |
| 0 | 0 | 1 | -0.4005 | 0 | 0 | | 0 |
| -1 | -1 | 0 | 1 | 0 | 0 | | 0 |
| 0 | 0 | 1 | -1 | 1 | 0 | | 0 |
| 0 | -1 | 0 | 0 | 0 | 1 | | 0 |

Det. A = 0.616529; Maximum error = 0.047092

| | Error | 0.02 | Error | 0.04 |
|---|---|---|---|---|
| | IC factor | 0.424698 | IC factor | 0.849396 |
| | Max | Min | Max | Min |
| dC | 0.141719 | -0.14172 | 0.910965 | -0.91097 |
| dI | 0.032149 | -0.03215 | 0.206653 | -0.20665 |
| dW1 | 0.116939 | -0.11694 | 0.751682 | -0.75168 |
| dY | 0.207645 | -0.20764 | 1.334738 | -1.33474 |
| dprofit | 0.124483 | -0.12448 | 0.800175 | -0.80018 |
| dK | 0.065926 | -0.06593 | 0.423773 | -0.42377 |

Error = 0.04, and Error = 0.02: Configuration for max all binaries: (A, b)

| | | | | | | | |
|---|---|---|---|---|---|---|---|
| 0 | 0 | 0 | 0 | 0 | 0 | | 1 |
| 0 | 0 | 0 | 0 | 0 | 0 | | 1 |
| 0 | 0 | 0 | 0 | 0 | 0 | | 1 |
| 0 | 0 | 0 | 0 | 0 | 0 | | 1 |
| 0 | 0 | 0 | 0 | 0 | 0 | | 1 |
| 0 | 0 | 0 | 0 | 0 | 0 | | 1 |

Configuration for min all binaries: (A, b)

| | | | | | | | |
|---|---|---|---|---|---|---|---|
| 0 | 0 | 0 | 0 | 0 | 0 | | 0 |
| 0 | 0 | 0 | 0 | 0 | 0 | | 0 |
| 0 | 0 | 0 | 0 | 0 | 0 | | 0 |
| 0 | 0 | 0 | 0 | 0 | 0 | | 0 |
| 0 | 0 | 0 | 0 | 0 | 0 | | 0 |
| 0 | 0 | 0 | 0 | 0 | 0 | | 0 |

**Exhibit 3.13**
**Uncovering Structures with Approximate Solutions**

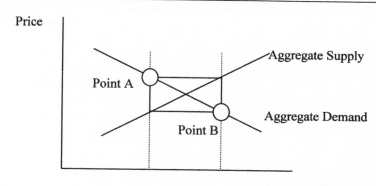

Observe now that the fluctuation allowed for an income change is larger than that caused by a fiscal policy change, even though the error we allow is roughly half of that which the model can take without breaking down. Adding errors allows us to obtain indicators of the quality of the model in practical settings where the errors will exist. In other words, presence of error creates a box where we know things happen but cannot further determine any more precisely what or how. If 0.01 is the error we attach to the model, then it is not possible to make inferences about policy effects, for example, without reducing that error further.

During the optimization both models exhibited internal consistency. For example, the configuration of binary variables that maximizes income, employment, and consumption is identical to that which minimizes price changes in the Keynesian model. This shows it is also possible to use the approximate model to obtain clues about the underlying economic structure. The maximum change in income that minimizes the change in prices is to be found moving downward along the aggregate demand curve. The maximum price change that minimizes income change will analogously move upward along the demand curve, as is shown in Exhibit 3.13. It is about what happens within the box of that picture that we cannot make reliable predictions.

In Klein's model, the configuration of binary variables that maximizes (minimizes), for example, the change in capital stock will maximize (minimize) other values as well. Economically, this indicates the consistency of the model in that an optimal capital expansion generates the best effects throughout the economy. Note that Klein's model utilizes real data. Mathematically, this may also indicate that the size of the errors overcomes the signal provided by the parameters and causes a situation in which the smallest inverse of A is always used to minimize (smallest b) or maximize (largest b).

It is appropriate to formulate some generalizations from the experience in building and solving approximate macroeconomic models. The reader should keep in mind that these observations are made in a state in which we know very little about the capabilities of the new models and may be too influenced by traditional modeling methods. Perhaps the best way to start these generalizations is by noting that the properties of the mathematical objects seem to blend with the theories and semantic interpretations of the problem. This is different from conventional modeling in which a matrix is simply a convenient object we apply in a variety of situations. In a way, traditional objects resemble hotel rooms and approximate equations houseboats in that the occupant's lifestyle is much more affected by the choice of dwelling. Let us provide some examples:

1.  It seems we can play around with wealth effects, rational behavior, dichotomy between the real and monetary sides of the economy, money neutrality, crowding out, Phillips curves, overshooting, and so on, only when we ignore errors. In contradistinction, in approximate equations we face a box where there may be room for all those "family pets" of economists.
2.  Modeling with intervals represents the expectations approach and nothing else comes closer. We think with approximations, and use intervals in everyday life. We are quite good at that: that is how we survive. In this sense, mathematical modeling corresponds more closely with the cognitive approach. AES help to clarify the role of expectations in economic models, which according to Kantor (1979) is one of the revitalizing factors of modern macroeconomics. Moreover, approximate equations are immune to Shackle's (1972) criticism of the probability approach, and Kuttner's (1985) critique of conventional macromodeling. It seems to us that, at the current state of knowledge, it is not possible to achieve the accuracy and reliability of knowledge that exact macromodels presume in terms of measurement, theoretical validity, individual behavior, and model specification. Further, exact macromodels seem to be optimal vehicle for endless "hair splitting" and untestable hypotheses.
3.  Another way in which to blend mathematical objects and theory is in considering model improvements and extensions. Testing alternative approximate models is easy: it amounts to calculating ill conditioning, errors, and comparing their respective "ignorance boxes." If the models have similar theoretical luggage and real data loads, the model with the smaller box would seem to be preferable.

From an operational point of view, modeling with approximate equations seems to blur some lines we are accustomed to take as given:

1.  The line between optimization and system description. In approximate equations, the links and variables themselves become "softer." Optimization is a process of refining and clarifying the potential roles and implications of those variables and relationships.
2.  The line between linear and nonlinear specifications is also erased, since both behaviors are possible within a box.
3.  The line between global and local behavior may vanish as well. Some errors may confine solution ranges to a "local" area, but other errors will make the box cover

very large portions of the parameter and solution spaces. When errors induce an ill-conditioning coefficient larger than one, the solution set may be insular, which means possible solutions may be scattered "all over the map."

4. Approximate equations also blur the lines between statics and dynamics since, again, within a box both behaviors are possible. A way to visualize this effect is to start with Samuelson's correspondence principle. Samuelson's (1983) correspondence principle says that a given solution of a static model contains the seeds of its dynamic behavior. In this sense, errors put into the structure of the model and the vector of exogenous factors induce a pseudo-dynamization.

5. The last of the lines blurred is that between stochastic versus nonstochastic models. Stochastic means pertaining or related to chance, and is generally confined to probability theory. It is easy to visualize situations in which the errors overcome the probabilistic residuals. In fact, econometric models that involve residuals are solvable, but certain approximation errors will break down the model. This issue makes reference to the adequacy of calculus versus calculus-plus-probability models to study practical problems. For example, Poole (1970) developed a model to study economic policies. However, his model becomes intractable when multiple stochastic behaviors are introduced, even though the model logically calls for precisely that. This means Poole's problem may be better studied via approximate equations.

Modeling with approximate equations also modifies our conception of the procedures we use for postoptimal analysis. As we have seen, it certainly limits the capacity of traditional comparative statics to predict what will happen to variables x, y, and z, when variable z1 changes. Approximate equations seem to represent an alternative to perturbation analysis, which is only valid for infinitesimal changes. A good way to start studying this issue is to posit the following question: How can one conduct comparative statics and perturbation analysis without considering errors?

Facing the presence of errors in a model changes the way we understand them. A policy maker or portfolio manager, obtaining advice from an economist might ask the economist to attach a likely error to the model to see whether her predictions and analysis have any numerical value at all or are simply feverishly concocted fiction. Considering the presence of errors requires that we do the following:

1. Review the relevant theory critically to select its most practical elements, which are likely to be those that stress the fundamentals.
2. Build models with common sense, which are manageable and can be explained to the users of such models.
3. Perform diagnostic tests to evaluate the fragility of the model via ill conditioning.
4. Attach an error to the coefficients and study the implied "ignorance ball" or "ignorance box" that we deal with.
5. Avoid any statements that cannot be numerically and reliably substantiated with the model numerics.

Imagine what a change! Current modeling procedures often stress central values, which neglect to indicate the uncertainty surrounding those values and the fact that we often focus precisely on extreme outcomes—as in the statement "Hope for the best, prepare for the worst." When some ranges are added, these are of a probabilistic nature and do not represent comprehensive testing—for example, they do not test for non-nested alternatives to the model. The reader may recall the many statements from economists, politicians, television announcers, radio people, and so on, about how changes in interest rates drove the stock market up and the exchange rate down. Or we may have heard those impressive explanations about why certain policies are better than others. Our response is always the same: Do we really know that much? Of course not! But if the issue is whether our current methodologies will set us on a more rewarding path, in this sense, approximate equations appear to be a very promising tool.

Whether it is possible to improve our knowledge by imposing accuracy or making explicit the lack of it is an epistemic issue we cannot address here. We certainly prefer having a model that tells us how much we do not know to a "glass museum piece" that hides our ignorance; but, of course, we have a bias for using models for practical purposes.

All in all, it is easy to conclude that approximate equation modeling is an area of macroeconomics that should be further explored.


## SUMMARY

This chapter presented the first application of approximate equations. We selected macroeconomics as the initial testing ground for approximate equations because many textbooks say that economic and financial—for example, asset allocation—decisions start with consideration of the economic environment. When we began our investigation, we did not know whether we would find interesting macromodels that could be made more practical via approximate equations. Our search for such models was thorough and very revealing. For example, we reviewed some of the most important models, from a theoretical point of view, and learned about the most important elements and relationships that should be preserved in any macroeconomic model. We also learned that developing a practical model might require a new start from the very foundations because the published literature does not offer models we could readily use with some degree of confidence in their usefulness.

Macroeconomic theory appears to offer considerable knowledge of what could be plausible economic variables and their potential relationships. However, economic theory cannot go very far without contrasting hypothesis empirically. Therein lie the major problems. Econometric models seem to add little knowledge to that provided on an a priori basis by the abstract formulations. Even authors who have developed models that have application potential do not bother to illustrate those models with "some numerics." This speaks volumes about their confidence in those models.

In general, it seems the goals of economists are heavily focused on economic policy issues, which stress the abilities of the model to explain and general questions such as efficiency of the price systems and effectiveness of economic policies to increase wealth and employment. This search is not, and may never be, effective. In the meantime, practical, low-cost macromodels that could be of help to individuals receive only a passing interest from the economics profession. In this study we have focused on practical decision-making and pursued development of that type of neglected model.

Above all, modeling with intervals presents a medium well suited to study economic expectations. This factor alone should make approximate reformulation of exact models attractive and desirable to the theoretically and empirically minded economist, especially in these times when we hear so much about expectational hypotheses.

At the very least, we may have shown it is indeed possible to develop low-cost macromodels with practical potential. It is important to note that when an individual in a firm develops such models, he or she is not responsible for "explaining the world" but simply needs to make sure a solid, professional effort has been applied to ground a particular business decision. In many cases, no one can claim to have a better forecast or understanding than that provided by the honest effort of the individual business decision-maker.

# 4

# A PRACTICAL-APPROXIMATE-FINANCIAL PLANNING MODEL

This chapter presents the second application of approximate equations.

The objective of this chapter is to develop a practical model and approximate methodology for financial planning in firms, which shall be done in three stages, each of which will correspond to the sections of this chapter. The first section reviews the basic elements of financial planning; namely, support from economic theory, problems and agents, financial management and planning, and alternative approaches to financial planning.

The second section concentrates on building a fundamental, generic, or basic financial planning model. It focuses on the core of financial planning—assessment of financing needs—and builds such a model from the algebra of financial statements. It shows that the simultaneous financial planning model obtained includes most important planning techniques (proforma statements), tools (break-even), and planning methods (sustainable growth, ROE analysis). This model is easy to understand and manipulate, and serves to highlight the most important characteristics of alternative financial planning strategies, especially for small firms for which external financing is not an option.

The third section adds errors to the model developed in the previous section and studies the implications of doing so. It starts by considering the sources of uncertainties in the financial planning problem, so that we can link the technique of adding errors to the SES model with the realities the model is supposed to represent. At this point it is sure that any practical model must be approximate, and we find the solution to the model and study its implications.

In developing this model, a strategy of simplicity for dealing with the complexity inherent in financial planning was followed. This strategy avoids normative pricing, simply because if we had a model that replicated actual pricing of stocks and bonds this book would not be necessary. Haugen (1996) provides additional reasons for not including normative pricing models in financial planning.

## 1. THE PROBLEM, AGENTS, AND THE MODEL

Our problem is to build a practical financial planning model for firms. The firm itself is an entity that maintains a balance, or equilibrium, between internal and external forces of a diverse nature such as productivity of inputs, technological development, sales, changing preferences, and so on. What do we learn from economic theory? What do we learn from financial management? How has financial planning been done up to now? These are the three questions we will try to answer in this section.

### A. The Support from Standard Economic Theory

In the previous chapter we implicitly defined firms and consumers as the production and consumption units of the economy, respectively. Firms carry out production and investment in capital goods. Unfortunately, the neoclassical-marginalist picture of firms and consumers developed in economics had serious limitations, especially when applied to financial planning. First, its construction was subordinated to create an aggregate model in which free markets and flexible prices were possible, and were conducive to economic equilibrium and prosperity (Walras' and Pareto's studies). From these models, economic research passed on to studying conditions in which freedom of choice in economic terms maps that of a political nature, as in the studies by Nobel laureates Kenneth Arrow, Kenneth Galbraith, and Milton Friedman.

Second, the micromodeling of consumers and producers was also subordinated to using the same mathematical objects (calculus). This makes things easy on the analyst but presupposes a precision and completeness we are far from having, as we have seen in the first and second chapters. Both consumers and producers are supposed to optimize a twice-differentiable function of preferences or costs subject to a linear budget constraint in a single-period problem, see Samuelson (1983), Chiang (1984), Henderson and Quandt (1980), and Varian (1984). There are many models that go beyond the simple static framework, but they all seem very experimental and speculative at this time. The next step up in terms of analysis is the state-preference model, which introduces contingent states, probabilities and interest rates, but which appears to simply formalize obvious and intuitive results. Analysis of stochastic economies (Markov-type, and Martingale-type) is studied in advanced texts but there is no pretense that these constructs depict any particular real world (see Huang and Liztenberger, 1988; Duffie, 1988; and Dothan, 1990). Interestingly, these probabilistic models often seem to bring to mind Shackle's (1972, p. 385) insightful observation that probability is only a model of thought, not a law of nature.

Third, the creation of aggregate equilibrium models was made possible by standardizing producers and consumers and by brutally simplifying economic decisions themselves and the context in which they are made. Consumers and producers are supposed to exhibit a homogeneity that is only plausible when strategic issues are ignored, or when preferences are similar, or when the analysis is kept within restrictive limits by, for instance, assuming choice between comparable goods or products lines, for example, detergent X versus detergent Y. The context is also simplified as well. It is appropriate to summarize some of those restrictive assumptions as follows:

1. The standard microeconomic theory of the firm assumes no inventory problems—Henderson and Quandt (1980) and Varian (1984)—while production-inventory issues are critical in practical business management.
2. Wages and labor are simplified to the extreme. In microeconomics we assume labor is homogenous, which is only reflective of what happens in certain areas of some manufacturing operations. In our modern firms, labor is basically human capital, not the "drones" implied by the neoclassical theory of the firm.
3. Capital stock is mostly machinery in the standard microeconomic model, for which we presumably can halt the rate of economic and technological obsolescence while we study issues concerning factor substitutability and scale.
4. Based on 2 and 3 above, we could say the production function and considerations used in microeconomics simply do not belong in any practical business setting.
5. Considerations regarding the cost of capital are collapsed into a single rate, presumably indicating what happens in a monetary (not financial) market, in a static setting. The existence of such a market must have predated the collapse of feudalism, if it ever existed at all, because straight barter probably played the role of financial markets (contingent commodity markets) even then.
6. Finally, the firm is assumed to operate in perfectly competitive markets—that is, it sells an undifferentiated product—while the majority of firms compete under conditions of imperfect competition, and the public sector may command about one third of the economic production/demand.
7. There is also the issue of the macroeconomics of corporate financial decisions. Cash balances, capital structure, and investing depend on expectations of macroeconomic variables. However, economics courses focus mainly on economic policy, which is not what investors and corporations care about most. As Jacob (1974) noted in the early 1970s, we need small, cost-efficient, forecasting models to assist investors in their asset allocation decisions. This discussion follows our presentation to the Academy of Economics and Finance entitled "Is Finance a Paper Tiger?" (Tarrazo, 1997b).

For us, it is not a question of forgetting that economics owes much of its origin to trying to show the superiority of the capitalistic system via the efficiency of its price system, but at some point we should pursue other issues, especially after the refreshing evidence provided by the collapse of the former Soviet Union and the unenviable economic and political conditions in China, or in any other collectivist society. Furthermore, progress in building practical models depends, in our opinion, on emancipating ourselves from the neoclassical straightjacket. First, by recognizing that all firms—as well as consumers—are vastly different. There are small and large firms, and firms with pricing power—due to monopolistic concessions, product differentiation, intellectual property, or competitive advantages. Second, by viewing firms, and consumers, as expectational units, as planning units for which every single action is made by thinking forward. It often seems we care only about the present, but perhaps this care is justified because the present is simply the cradle from which we expect an attractive future to emerge.

It turns out that the main lessons to be learned from a course on managerial economics are mostly qualitative, not quantitative, ideas, but they are very important. The following is our short list:

1. Decreasing marginal productivity when at least one factor is fixed. This idea limits options concerning factor substitutability. In general, firms face some choice when selecting labor, capital, and sometimes location and technology. Still, when a factor of production is fixed, the bottleneck will be felt sooner or later.
2. Optimal capacity. Every productive activity has an optimal capacity, which is often related to the size of the operation and propitiated by the optimal input mix. We cannot assess this optimal size-input mix because everything is constantly changing and because firms are not biological entities with biological constraints; but profit optimization is still an endless fine tuning of input mixes and output sizes.
3. Barriers to entry make competing via price very difficult for new entrants. This is a fundamental lesson that is critically important for writers of business plans for new ventures: If you have a small firm, you will face high average costs and will not be able to offer higher quality at a lower price than the competition—unless you have something proprietary (a patent, government license, and so on).
4. Product differentiation is one way to obtain pricing power. The way a small firm, with an output most likely short of optimal capacity, breaks into a new market is through product differentiation, perhaps associated but not necessarily so to higher quality.

With these fundamental ideas from microeconomics in mind, we may ask ourselves a simple question: Why do firms exist? Alchian and Demsetz (1972) suggested that team production—not long-term contracts between employers and employees—is the essence of the organization we call the firm. According to these authors, team production is when: (1) several types of resources are used; (2) the product is not a sum of separable outputs of each cooperating resource; and (3) not all resources used belong to one person (Alchian and Demsetz, 1972, p. 779). Malmgren (1961) gave a better answer. He views the firm as a planning or expectational unit that could reduce some of the external uncertainty in its operations by bringing resources under the "umbrella" of the firm. Still, there is much we need to know about the basic nature of the firm in order to properly evaluate issues related to mergers and acquisitions, and the provision of goods and services by the public (federal, state, and local government) sector. The reader may find these questions too speculative; more practical questions are also in need of practical answers. For example, what do firms maximize? What is the correct size for a given business venture? However, we do not want to get lost in metaphysics of the firm. Viewing firms as planning units is the major step forward we need to build practical models.

## B. Financial Management and Planning

The next step in our collection of materials with which to build a practical model is, of course, financial management. Textbooks on financial management or corporate finance for undergraduate, graduate, or executive students present material that is disjointed. According to most textbooks, financial management seems to take two routes. On one hand we have a series of "components"—time value of money, working capital, capital budgeting, and capital structure—that constitute most of what students learn in introductory courses. Textbook authors emphasize what they call "modularity," but this is the best way to express the disconnected character of each of these parts. That is, working capital management is deficiently linked to capital budgeting or capital structure. In sum, when we learn financial management from each of these modules, we are basically "planning by parts." (Working capital management refers to managing short-term assets such as cash, inventories, accounts receivable, and accounts payable. Managers should not match short-term assets to short-term liabilities, as is said in some texts, but predictable—short and long term—financing needs to long-term financing and use short-term financing to cover unpredictable needs beyond those accounted for in cash budget planning. This is an important link often missed in both theory and practice of financial management.)

It is agreed that what is important in corporate decision-making is to manage by planning and not by crisis. Planning becomes more important when the planner has little control over information in the external or internal environments. In this sense, planning is most important for small firms. According to Fry, small businesses and entrepreneurial ventures fail not due to lack of resources, but to lack of planning (1992, Chapter 1). In one of the few textbooks emphasizing financial management from a small-firm perspective, Petty et al. (1993) provide the following specific reasons explaining the importance of planning for small firms: (a) they are less liquid and carry lower working capital management ratios, which makes short-term cash flow patterns critical, (b) they have more volatile profits and cash flows, (c) they use relatively large amounts of debt, perhaps because they do not have access to financial markets, (d) they lack access to capital markets making estimation of cost of capital and firm valuation very difficult, and (e) they rely on internal financing as evidenced by their high retention ratios. Corman and Lussier (1996) have emphasized a planning approach to small firm management recently, in addition to Fry (1992), and Petty et al. (1993).

In a general sense, finance is concerned with how individuals and firms allocate resources over time, Fama and Miller (1972). This definition emphasizes how finance is concerned with physical, as well as financial, resources. The focus on intertemporal decisions implies the development of quantitative methods to deal with uncertainty. Finance, we could say, is about planning. Therefore, one would expect this subject to be central in financial management textbooks. Yet, even premiere textbooks such as Brigham and

Gapenski (1993) or Gitman (1993), usually allocate no more than a couple of chapters to the topic.

Financial planning is studied in four stages. First, as we have noted, we learn about planning by parts (cash, inventory, accounts receivable, capital budgeting, and so on). Second, we employ dedicated procedures or methods (percentage of sales method for proforma statements, break-even analysis, EBIT-EPS, and so on) to study particular questions about planning. Third, we develop frameworks of analysis such as sustainable growth, return-on-equity (ROE) or DuPont analysis, and cash-flow analysis. The fourth stage of planning is the comprehensive approach, which is the least known of all. We can distinguish two approaches within this fourth stage: the simultaneous equations model of Warren and Shelton (1971) and the (linear) programming model of Carleton (1970) and Carleton et al. (1973).

There are only two financial planning models proper: linear programming and simultaneous equations, which we shall briefly review next.

## C. Alternative Approaches

Warren and Shelton (1971) (W&S) and Francis and Rowell (1978) present the algebraic or simultaneous equation (SES) approach to financial planning, which is particularly relevant for the purpose of this monograph.

W&S organized their model into four sections: (1) generation of sales and earnings before interest and taxes, (2) generation of total assets, (3) financing the desired level of assets, and (4) generation of per share data. Some of these equations can be subsumed or included into others via variable substitution. From these four segments, the only ones that need comment are the third and fourth, which deal with debt and equity, respectively. In the third section we can observe equations dedicated to debt amortization, interest payments on debt liabilities, and the actual leverage ratio—equations (9), (12), and (13), respectively, in W&S model. These equations (and the rest in their fourth block) reveal that the model was designed mainly to generate multiperiod proforma statements. On a period-by-period basis, one would not need, nor use, automatic "carry-over" equations because management may wish to change some of these items, for example, the leverage ratio. This is not a limitation of the model. When we build multiperiod projections in one pass we must, obviously, give up changing things interactively. Our objection, however, is that we know better than that. There are methods of planning such as the sustainable growth rate that can be inserted into automatic multiperiod projections to provide some corrective feedback in the model.

The fourth block of equations is more problematic from a theoretical viewpoint, because they depend on a particular pricing model—the price-earnings-ratio (PER) model, equation (18)—whose performance or adequacy as a pricing instrument is far from being established. If, on one hand, W&S include this model to build one-pass multiperiod projections, we can understand its inclusion as long as the projection period is not so long that market perceptions concerning the optimal PER might change. It is not clear whether "mt" is the historical or optimal one, which is a very major issue. (Fisher Black, 1980, p. 6, has argued that analysis of "fundamentals" for investments purposes may center on finding divergences between historical and optimal PER's, which also suggests that management policies would aim to correct such divergence.) If, on the other hand, the PER model is included as a true representative of a valid and acceptable class of pricing models we would have strong objections: We do not know of any single pricing model with an established record of performance, and we seem as close to finding such a model as alchemists were to finding plogiston a few hundred years ago. As noted later in the chapter, within the context of calculating cost of capital and financing, Haugen has recently written that "given the current state of the field, this technology should not be based on theoretical models" (1996, p. 96), a statement that also applies to financial planning.

Francis and Rowell's analysis contains greater detail (36 equations) than Warren and Shelton's (20 equations). Burrill and Quinto (1972) present an even more detailed example of simultaneous equations. The problem with more equations is both the assumptions and implicit connections they imply, as well as the presumed accuracy.

Carleton (1970) first proposed the programming approach to financial planning. The methodological aspects of the programming approach are discussed in Carleton, Dick, and Downes (1973), later by McInness and Carleton (1982), and by Tarrazo and Gutierrez (1997, 2000). This approach was designed to overcome some of the limitations of the SES approach, namely, no optimization, avoidance of difficult choices since they only trace out the consequences of decisions, incorporation of more accounting than finance, and improper use of equalities when inequalities are more relevant.

Carleton's optimization model (see Tarrazo and Gutierrez, 1997) is composed of an objective function that maximizes the firm's share price subject to eight constraints that can be grouped in three categories—definitional, sources and uses of funds, and policy constraints. As discussed earlier, such a model attempts to integrate both investment and financing variables, for this linkage is a key issue in financial planning. This integration results in a restrictive model that not only asks for numerous inputs from the manager but also incorporates more uncertainty into the process. Therefore, its applicability is limited to—increasingly fewer—companies in fairly stable industries. Most firms, however, are exposed to changing environments characterized by a great deal of uncertainty, thus requiring a more flexible model. In Tarrazo and Gutierrez (1997), we simplified Carleton's original model and applied fuzzy mathematical programming to the simplified version.

The simplified model introduced in Tarrazo and Gutierrez (1997) consists of an objective function and four constraints: definitional (available-for-common-funds, or AFC), equal sources and uses of funds (or SUF), maximum leverage, and payout restrictions. By eliminating three of Carleton's policy constraints, the manager does not have to input variables such as the estimated P/E ratio at the end of the planning horizon. She also does not have to determine the expected stock price growth rate or the post-horizon dividend-growth rate as required by Carleton's model.

We have already criticized some areas of current financial modeling for planning purposes (for example, economic support, lack of integration among financial management modules, and so on). The historical simultaneous equation and linear programming models can be criticized on several grounds as well, all of which center on how they deal with uncertainty. Instead of leaving room for uncertainty within the model, the usual procedure is to make assumptions to remove it, which cripples models for any practical purpose. For example, in Warren and Shelton's (1971) model, we remember that two of the variables provided by management—expected interest rate on new debt and the price earnings ratio—link the inner world of the firm and its financing problem to the external environment of debt and equity markets. The addition of pricing segments to the basic financing problem removes the uncertainty concerning how the market values the firm but also precludes us from facing this uncertainty and learning and developing alternative models.

Uncertainty is removed from the picture in Carleton's model in several ways. For example, the model requires the planner to input critical data such as the appropriate discount rate or required return on equity, expected earnings before taxes for the coming and end-of-planning-horizon years, investment in capital assets and annual profits from such investments, expected interest rates for long-term debt during the planning horizon, and internal rate-of-return per period earned on the investment (gross profit margin is used as a proxy), and so on. Much of the impressive detail of Carleton's model is achieved through assumptions and decisions that remove the three major sources of uncertainty in financial planning: sales growth, cost of new equity, and cost of new debt. Carleton himself recognized the limitations of his approach: "One of the principal advantages to the present model would seem to be in fact that it clearly reveals the major modeling to be done" (1970, p. 653).

As we noted in Tarrazo and Gutierrez (1997), traditional models are too detailed and too normative, which makes them impractical. Moreover, the model becomes very fragile to criticism; and, more importantly, greater detail is obtained at the cost of eliminating uncertainty: "The order-of-magnitude increase in model complexity occasioned by formal introduction of uncertainty would severely limit its acceptance at the management levels at which a financial systems' frame of reference has greater value" (Carleton, 1970, p. 653). But we have to remember the following: Uncertainty is the only reason we engage in financial planning. The only strategy for including uncertainty in the model and making it practical is to simplify the model. Finally, obsessive emphasis on normative modeling leads to situations that would appear ridiculous to any practical person. For example, some working capital planning models assume the manager knows the exact evolution of long and short term interest rates, and the GNP. Both the simultaneous equation and programming approaches available in the literature assume the planner knows how the stock market works and how security prices are determined. Knight (1972) went further in his appraisal of programming models and proposed that the role of optimization models in financial planning be severely limited.

In Tarrazo and Gutierrez (1997) we found the SES approach better suited to evaluate internal financing policies given environmental stability, and the programming approach useful for evaluating optimal responses to changing environments. This means these approaches are not competing, but complementary. Our studies have also agreed with Malmgren (1961, p. 405) that firms are interested in pursuing policies conducive to stability. Developing reliance on internal financing, as is shown in the remainder of this chapter, can attain this stability. It is well known that financial self-reliance is higher the larger the sales growth and profit margin of the firm, and is also conducive to maximizing the value of the firm. Sun Microsystems and Microsoft Corporation are suitable examples of corporations that historically have relied on internal financing strength.

The next section of this chapter develops a practical financial planning model. It was initially designed to meet the planning needs of small to medium-sized firms, but is equally appropriate for large firms. It is based on an algebraic treatment of the basic financial statements of the firm, and encapsulates popular planning methods such as proforma statements, break-even, and DuPont (ROE) analyses. Uncertainty is introduced by viewing the simultaneous equation system as an approximation.

The core problem that drives other business planning decisions is that of determining financing needs, which depend on working capital management and capital budgeting, and past decisions concerning the structure of liabilities of the firm.

## 2. THE FOUNDATIONS OF A PRACTICAL FINANCIAL PLANNING MODEL

The objective of financial planning is to identify and secure the necessary financing to support sales estimates. This objective is central to all planning tools and methods. The first segment of this section presents a model built from the basic financial statements of the firm, which makes it practical and easy to understand. The second segment shows how the model we present incorporates popular planning methods and techniques.

### A. The Algebra of Financial Statements

Financial statements are effective instruments for assessing financing needs. Sales growth will often be accompanied by growth in fixed and current assets, which must be financed through internal (retention) and/or external sources (new issues of debt and equity). As we know, accounts payable may provide additional financing, but this amount will not be enough to support sales growth by itself. We can assess financing needs by examining prospective income, balance, and flow-of-funds statements.

Let the following variables represent the financial statements of the firm for two given periods:

Rnew =  retained earnings for the next period
b     =  retention rate
tax   =  effective corporate tax rate
EBIT  =  earnings before interest and taxes
iold  =  interest rate on previous period debt
Dold  =  existing longterm debt
inew  =  interest rate on new debt
Dnew  =  new long term debt
Ul    =  flotation costs of new debt
Us    =  issuing costs of new equity
Prefdiv = preferred dividends
FN    =  financing needs
TA    =  total assets
CL    =  current liabilities
Rold  =  old retained earnings
PREFS =  preferred stock
RCA   =  current assets to sales ratio
RFA   =  fixed assets to sales ratio
RCL   =  current liabilities to sales ratio
Salesnew = expected next period sales = Salesold (1 + sales growth rate)
w     =  leverage ratio

The income statement can be written as

(1)          Rnew =
   = b{(1-tax) [EBIT - iold Dold - (inew+Ul) Dnew - Us Snew] - Prefdiv}

The flow-of-funds or cash-flow statement, which links the income and balance statements, can be written as

(2)          FN = TA - CL - Dold - Sold - Rold – Rnew =
   = (RCA + RFA - RCL) SALESnew - Dold - Sold - Rold   - PREFS – Rnew

External financing is required when new retentions are insufficient to cover the net growth in assets. We are assuming the amount of funding needed goes beyond that achievable through short-term sources such as lines of credit or short-term loans.

(3)          FN = Dnew + Snew

At this point, we have three equations but four unknowns, which are the financing required (FN), amount to be retained (Rnew), new equity (SNew), and new debt (DNew). A fourth equation such as

(4)          Dnew = w SNew

can be added to complete the system of simultaneous equations. Exhibit 4.1 shows the results of rearranging the equations to form the customary, A x = B, matrix operands. The matrix A is invertible under very mild conditions, which ensures a unique optimal solution to the financing problem. (Expansion by alien cofactors of matrix A using, for example, the first column shows the conditions under which its determinant will not be equal to zero. The condition is $w > (1 + a2)/(a2 - a1 - 1)$. For the non-levered firm, w=0, it amounts to $a2 > -1$, or $b(1-tax)$ Us $> -1$, which is very easily met with normal ranges for the variables involved.)

**Exhibit 4.1**
**A Simple SES Model for the Financing Problem**

Rnew + [b(1-tax)(inew+ Ul)] Dnew + [b(1-tax)Us] Snew     + 0  = C1
Rnew +                              +                     + FN = C2
 |  0  + Dnew                    |  + Snew    |           - FN =| 0
 |  0  + Dnew                    |  - w Snew  |           - 0 = | 0

| 1  | a1 | a2 | 0  | | Rnew |   | | c1 |
| 1  | 0  | 0  | 1  | | Dnew |   | | c2 |
| 0  | 1  | 1  | -1 | | SNew | = | | 0  |
| 0  | 1  | -w | 0  | | FN   |   | | 0  |

a1 = [b(1-tax)(inew+ Ul)]
a2 = [b(1-tax)Us]
c1 = b{(1-tax)EBIT}SALESnew + k1
   = b (1-tax) REBIT SALESold (1 + g) + k1
k1 = - b{[(1-tax) iold Dold] - Prefdiv}
c2 = (RCA + RFA - RCL) SALESnew
   = (RCA + RFA - RCL) SALESold (1 + g) + k2
k2 = - Dold - Sold - Rold - PREFS
g  = expected growth rate for sales

It is not redundant to note that whatever we do in financial planning must conform to the system of four simultaneous equations, for they are the financial statements of the firm. Equation (1) shows the flow of funds in the income statement and the residual nature of retained earnings. Equation (2) shows the flow of funds through the balance sheet. Divergence in the assets-liabilities must be met by either new issues of debt and equity, or by internal funds. This is the information provided by equation (3). Equation (4) says that new debt offerings must be proportional to new stock offerings. This policy could be changed without creating any problem for the system of equations.

The four-equation model has a number of interesting features.

1. It provides financing needs in terms of dollar values. For example, the model says the firm will need $200,000 in equity financing whatever its stock price may be. This characteristic of the model is appropriate for privately held firms without access to organized debt and capital markets. This feature is also handy when considering the capital rationing conditions under which most firms operate, especially small and entrepreneurial firms. (Carleton criticized the simultaneous equation approach on the basis of utilizing "accounting rather than finance," 1973, p. 565. However, finance is more than optimizations and normative pricing models. In our opinion, finance is about making quantitative decisions under uncertainty. Any model suitable for that purpose is a financial model, especially if it helps to solve practical problems.)
2. As we will show later in this chapter, this model permits the calculation of sales growth that can be financed internally. The no-external-financing growth rate is compatible with the finding that firms are reluctant to issue new stock, and is also a useful concept for small firms for the reasons mentioned in the previous paragraph.
3. The four-equation model can incorporate uncertainty because it is simple. Not only does our model prevent us from assuming we know more than we actually do (normative pricing models), but it makes it easier to study optimal responses to changing conditions. It is reasonable to speculate that emphasis on normative pricing issues was central during the 1960s, which were characterized by relatively stable economic growth and less complex financial markets and economies. Nowadays, it may be appropriate to shift the emphasis onto uncertainty—real uncertainty, not the kind we presume to handle with probability theory—given the complexity of our markets and the (justified or not) anxiety seemingly prevalent in our economies.

Our elimination of normative pricing equations from the model may be regarded as a serious departure from conventional financial modeling. However, Professor Haugen has recently written, "given the current state of the field, this technology (to calculate cost of capital and financing) should not be based on theoretical models" (1996, p. 96).

Exhibit 4.1 presents the model in a matrix form, which is very convenient for spreadsheet manipulation. Exhibit 4.2 presents an application of the four-equation model. The content of Exhibit 4.2 mimics the spreadsheet model used in the calculations. The example is self-explanatory. The firm targets a growth rate for sales of 10%, which would require issuing 44.95 units of new debt and 89.91 units of new equity. The dollar amount of new debt and new equity equals the value of FN (financial need), which also equals the difference between internal funds and (assets-liabilities) growth.

**Exhibit 4.2**
**The Simplified Model. An Example**

Data

| | | | |
|---|---|---|---|
| Iold | = 0.05 | Stock price | = 2 |
| Inew | = 0.1 | Sales (t=0) | = 500 |
| Ul | = 0.005 | b | = 0.5 |
| Us | = 0.005 | w | = 0.5 |
| Tax | = 0.35 | g | = 0.1 |
| Prefdiv | = 0.08 | | |
| REBIT | = 0.8 | | |

Income Statement (t=0)

| | |
|---|---|
| SALES | = 500 |
| EBIT | = 400 |
| INTEREST | = -37.5 |
| U-DEBT | = 0 |
| U-STOCK | = 0 |
| EBT | = 362.5 |
| EAT | = 235.625 |
| PRDIV | = -8 |
| EARNINGS | = 227.625 |

(0.8 next to EBIT)

Balance Statement (t=0)

| | | Ratios | |
|---|---|---|---|
| CA | 1000 | 2 | = RCA |
| FA | 2000 | 4 | = RFA |
| Sum | 3000 | 6 | = RA |
| CL | 400 | 0.8 | = RCL |
| DOLD | 750 | 0.4583 | = w |
| SOLD | 1636.1875 | | |
| PREF | 100 | | |
| ROLD | 113.8125 | | |
| Sum | 3000 | | |

(0.8 next to FA)

Coefficients

| | |
|---|---|
| a1 = | 0.034125 |
| a2 = | 0.001625 |
| k2 = | -2600 |
| k1 = | -16.1875 |
| c2 = | 2860 |
| c1 = | 143 |

DuPont Analysis

| | |
|---|---|
| PM | = 0.47125 |
| AT | = 0.1666667 |
| ROA | = 0.0785417 |
| Eq. Mult | = 1.8335307 |
| ROE | = 0.1440086 |

$$
\begin{vmatrix}
1 & 0.034125 & 0.001625 & 0 \\
1 & 0 & 0 & 1 \\
0 & 1 & 1 & -1 \\
0 & 1 & -0.5 & 0
\end{vmatrix}
=
\begin{vmatrix}
126.8125 \\
260 \\
0 \\
0
\end{vmatrix}
$$

$$
\begin{vmatrix}
1.0126155 & -0.0126155 & -0.0126155 & -0.02194 \\
-0.3375385 & 0.3375385 & 0.3375385 & 0.67398 \\
-0.675077 & 0.675077 & 0.675077 & -0.65204 \\
-1.0126155 & 1.0126155 & 0.0126155 & 0.02194
\end{vmatrix}
=
\begin{vmatrix}
125.132 & \text{Rnew} \\
44.9559 & \text{Dnew} \\
89.9118 & \text{Snew} \\
134.8673 & \text{FN}
\end{vmatrix}
$$

**Exhibit 4.2 (continued)**

| Income Statement (t+1) | | Balance Statement (t+1) | | | |
|---|---|---|---|---|---|
| | | Ratios | | | Ratios |
| SALES | = 550 | CA | 1100 | 2.2 | = RCA |
| EBIT | = 440 | FA | 2200 | 4.4 | = RFA |
| INTEREST | = -41.995591 | Sum | 3300 | 6.6 | = RA |
| U-DEBT | = -0.2247795 | CL | 440 | 0.88 | = RCL |
| U-STOCK | = -0.4495591 | DEBT | 794.955 0.460 | = w | |
| EBT | = 397.33007 | STOCK | 1726.0993 | | |
| EAT | = 258.26455 | PREF | 100 | | |
| PRDIV | = -8 | RETS. | 238.94477 | | |
| EARNINGS | = 250.26455 | Sum | 3300 | | |

| Sustainable Growth | | DuPont Analysis | |
|---|---|---|---|
| d1 = | 130 | PM | = 0.4695719 |
| d2 = | 2600 | AT | = 0.1666667 |
| | | ROA | = 0.078262 |
| g = | 0.0460779 | Eq. Mult | = 1.9118251 |
| | | ROE | = 0.1496232 |

| Old Market Data | | New Market Data | |
|---|---|---|---|
| Stock price | = 2 | 100 | = new shares issued |
| Number of shares | = 818.09375 | 918.09375 | = total shares |
| | | 1.8800905 | = Price new |
| EPS | = 0.2782383 | 0.2725915 | = EPS |
| PER | = 7.1880835 | 6.8970989 | = New PER |

Retentions check
b*earnings 125.13227
rold      113.8125
Sum       238.94477
RETS.     -238.94477 (125.1327 Rnew, from matrix inversion, plus ROLD.)
Difference   0

---

Calculation of market data requires that we provide (at least) the number of new shares issued. The new price for common stock can be calculated by dividing the value of equity by the total number of shares. By changing the values for b and w—retention and original debt/common stock equity, respectively—we can study the basics of alternative financing policies. However, the practical usefulness of the market data calculated in this manner is limited by the lack of an effective ex-ante pricing model.

## B. Other Planning Tools and Methods

As noted in the introduction, financial planning is made up of planning by parts, tools, and methods. It is easy to show that the model we propose is inclusive of popular planning tools such as proforma statements and break-even analysis, and also includes planning methods such as ROE analysis and sustainable growth.

DuPont analysis is subsumed into our equation system because the system contains all information the DuPont system handles. Return on equity is the relevant variable in DuPont analysis: ROE = Profit margin x Asset turnover x Financial leverage = (Net Income/Sales) x (Sales/Assets) x Assets/Equity).

As was the case with DuPont analysis, our equations also contain other financial planning tools such as proforma statements, cash-flow analysis, and sustainable growth assessment. The sustainable growth rate is the growth in sales compatible with current policies and the parameter of the firm and implies using new, external financing (see Higgins ,1977, and 1995; and Brigham and Gapenski, 1993, p. 738). (As Higgins (1995, p. 125) notes, the sustainable growth rate could be calculated as g* = b x ROE, which requires us to redefine "equity" as beginning-of-period equity. Brigham and Gapenski (1993, p. 738) calculate the sustainable growth rate using the following equation: g* = [M b (1+(D/E))]/[(A/S)—M b (1+(D/E))]. In this specification, M represents the projected profit margin and assumes the ratios will not change in the coming year. D, E, A, S, and b are debt, equity, assets, sales, and retention ratio, respectively.)

Exhibit 4.3 presents a numerical example in which all data concerning alternative planning tools has been calculated via the four-equation model. In addition to the standard sustainable growth model, our model permits calculation of the no external financing growth rate, whose importance for small firms will be discussed in the next section.

## C. Delineating Financial Strategies

As a decision-making model, a financial planning model should be a helpful tool to learn about planning and assess contingent strategies. Capability to finance corporate growth, and the internal or external origin of this financing, are the two most critical issues concerning planning. This section will show how the four-equation model sheds light on these important issues.

### 1. The Quest for Self-Sustained Growth

The no external-financing growth rate (g*) can be calculated as follows:

$$(5) \qquad g^* = \frac{[d2 - d1 + k2 - k1]}{[d1 - d2]}$$

**Exhibit 4.3**
**SES and the Financing Problem: Sustainable Growth**

$$g = 0.0460779$$

Coefficients
$a1 = 0.034125$
$a2 = 0.001625$
$k2 = -2600$
$k1 = -16.1875$
$c2 = 2719.8026$
$c1 = 135.99013$

$$
\begin{vmatrix}
1 & 0.034125 & 0.001625 & 0 \\
1 & 0 & 0 & 1 \\
0 & 1 & 1 & -1 \\
0 & 1 & -0.5 & 0
\end{vmatrix}
=
\begin{vmatrix}
119.80263 \\
119.80263 \\
0 \\
0
\end{vmatrix}
$$

$$
\begin{vmatrix}
1.01 & -0.01 & -0.01 & -0.02 \\
-0.34 & 0.34 & 0.34 & 0.67 \\
-0.68 & 0.68 & 0.68 & -0.65 \\
-1.01 & 1.01 & 0.01 & 0.02
\end{vmatrix}
=
\begin{vmatrix}
119.80 & \text{Rnew} \\
0 & \text{Dnew} \\
0 & \text{Snew} \\
0 & \text{FN}
\end{vmatrix}
$$

Income Statement (t+1)

| | |
|---|---|
| SALES | = 523.03897 |
| EBIT | = 418.43117 |
| INTEREST | = -37.5 |
| U-DEBT | = 0 |
| U-STOCK | = 0 |
| EBT | = 380.93117 |
| EAT | = 247.60526 |
| PRDIV | = -8 |
| EARNINGS | = 239.60526 |

Balance Statement (t+1)

| | | Ratios | |
|---|---|---|---|
| CA | 1046.0779 | 2.0921559 | RCA |
| FA | 2092.1559 | 4.1843117 | RFA |
| Sum | 3138.2338 | 6.2764676 | RA |
| CL | 418.43117 | 0.8368623 | RCL |
| DEBT | 750 | 0.4583827 | w |
| STOCK | 1636.1875 | | |
| PREF | 100 | | |
| RETS. | 233.61513 | | |
| Sum | 3138.2338 | | |

This rate is obtained from the flow-of-funds condition noting that d1 (1+g) + k1 = d2 (1 + g) + k2, where d1 = b(1 - tax) REBIT Salesold, and d2 = (RCA + RFA - RCL) Salesold.

Exhibit 4.3 shows the numbers obtained for the no external financing maximum growth rate (4.60%). This number is calculated by equation (5), and is identical to the number obtainable by using the "goal seek" facility in the spreadsheet for FN = 0.

Strengthening the no external financing growth rate is perhaps the most fundamental way to enhance the health of the firm. Strengthening this rate on a period-by-period basis is also the best way to increase the market value of the firm. Note that the growth rate in sales, earnings, or cash flow is the key variable in both the dividend growth model and the price-to-earnings valuation model, barring external effects. Gordon's constant growth dividend model can be rewritten as $ki = dy + g$, where $ki$ is the investor's required rate of return on equity, $dy$ is the dividend yield, and $g$ is the growth rate in sales, earnings, or cash-flow. The price-to-earnings ratio model can be written as $pt = pert * epst$, where $pt$ is the stock price, $pert$ is the (optimal) price-to-earnings ratio used by the market, and $epst$ is the earnings per share of the firm during the period of reference. Obviously, maximizing $g*$ also maximizes $epst$ and $pt$. Moreover, it is hard to imagine why anyone would lend money or invest in a firm that will not experience sales growth.

Including uncertainty in the financial planning problem requires that we assess the funding capabilities of the firm. Note that reliance on internal financing is critical for small firms, particularly in countries with underdeveloped financial markets.

The prescription that has emerged from our analysis is to strengthen the no external financing growth rate, which is consistent with Petty et al. (1993) finding of low pay-out (high retention) ratios for small firms. While the firm is doing this, it will have come up with optimal responses to changing external conditions. The development of such policies is studied next.

## 2. Optimal Responses and Small Firms

The four-equation model could have been solved in a number of ways. Presence of a few variables in the right-hand-side vector B of the system of simultaneous equations was not an accident, but a modeling strategy to study optimal responses. These responses will try to neutralize falls in growth of expected sales.

Observing the equations, we notice that sales variations are transmitted through a set of ratios, where $k1$ and $k2$ are constants representing historical values:

(6)      $c1 = b\{(1\text{-tax})EBIT\}SALESnew + k1$
          $= b\,(1\text{-tax})\,REBIT\,SALESold\,(1 + g) + k1$

(7)      $c2 = (RCA + RFA - RCL)\,SALESnew$
          $= (RCA + RFA - RCL)\,SALESold\,(1 + g) + k2$

The variable $c2$ links the effects of sales variation to the balance sheet, while $c1$ does likewise to the income statement. The implication is that changes in these ratios would amplify or dampen the effects of sales variability of the firm. Moreover, $c1$ and $c2$ show why a profitable firm can go bankrupt: net income may not pay for the growth in assets required by sales. In milder cases, as Higgins notes, rapid growth puts companies on a treadmill: "The faster they grow the more cash they need, even if they are profitable" (1995, p. 126).

According to the valuable research by Petty et al. (1993), small firms have the following financial characteristics:

1.  High retention ratio (b).
2.  Low working capital ratios (RCA, large RCL).
3.  Low liquidity—short-term cash flow patterns are necessary.
4.  High business risk and more volatile profits and cash flows.
5.  Use of relatively large amounts of debt, perhaps because they do not have access to financial markets.
6.  Lack of access to capital markets, making estimation of cost of capital and firm valuation very difficult.

In these conditions, reducing the impact of sales fluctuations will be very difficult. For example, "managing" c2 would require firms to lower working capital ratios to neutralize the fall in sales; but doing so is very difficult because small firms already have these ratios at minimal values. In passing, note how well the model explains characteristic (4) above. Similarly, dealing with c1 amounts to increasing the efficiency of the firm (higher REBIT), since retention rates are at a maximum and taxes should be at a minimum.

The implication is that managing downturns in sales requires solutions beyond those of a purely financial nature. This is well known to small business managers but not so patent in financial management textbooks. Optimal responses to low growth require the integration of financial policies, as well as others. For example, small firms may pursue capital budgeting strategies conducive to developing "niches" in order to develop stability in sales revenues. Niches are likely to be associated with high product prices. However, this should not be a source of problems because small firms are unlikely to compete with large firms via price, since they cannot reach scale economies. Finally, another element not generally studied in financial management, strategic analysis of growth opportunities, is likely to be critical in managing small firms by planning and not by crisis.

## 2. THE PRACTICAL FINANCIAL PLANNING MODEL

For both large and small firms, existence is predicated on growth and renewal. Both of these require planning. The related issue of optimal size is one of the questions that economics and business research strive to answer. Malmgren's (1961) study of the firm as a planner notes that firms may get larger to internalize prices and variables that would otherwise be beyond their control. According to Malmgren's expectation-based analysis, firms' policies are designed to provide stability. This section, therefore, will extend Malmgren's analysis by examining the sources of uncertainty for firms and ascertaining optimal responses (in Malmgren's sense) to changing conditions. Uncertainty is introduced via approximate equations systems, which permit solutions in interval form. There are other ways to incorporate uncertainty into a financial planning model. For example, Tarrazo and Gutierrez (1997) apply fuzzy mathematical programming to obtain solutions in interval form, and Tarrazo (1997i) presents a fuzzy-set based methodology and model for qualitative planning.

## A. Sources of Uncertainty

The introduction noted that financial models are not constructed to deal with uncertainty, even though uncertainty is the reason financial management exists. In financial planning, there is uncertainty concerning internal factors of the firm such as productivity, operating efficiency (REBIT), and financial efficiency. There is also uncertainty regarding the external environment.   Exhibit 4.4 presents the information flows and relationships relevant to the planning decision.

For small firms, the main sources of uncertainty are 1, 2, and 3, which relate to sales expectations and financing conditions.  Financial planning concerns how to finance expected growth in sales.  For this purpose, Higgins' sustainable growth model mentioned in the previous section is a useful tool when firms have access to capital markets.  The concept of maximum affordable growth with no external financing is relevant for small firms and entrepreneurial ventures, which do not have access to external financing.  Even large firms in well-developed capital markets have been reluctant to raise new equity, or have found it very difficult.  Higgins (1995) provides statistics indicating that internal sources, depreciation, and increases in retained earnings accounted for 65% of total financing for U.S. nonfinancial corporations during the 1965–1993 period. Higgins notes that American corporations retired more stock over this period than they issued, and summarizes potential explanations as follows: (a) companies may not need financing beyond retentions and new profits, (b) equity is expensive to issue, (c) managers may not want to dilute earnings-per-share, (d) managers may believe their stock is undervalued, and (e) managers may perceive the stock market as an unreliable source of funding.

Decision-making models such as the one presented have the following elements: (a) a collection of mathematical objects (algebraic equations and matrices in our case), (b) theories that provide meaning to the model (flow of funds, accounting statements, sustainable growth, DuPont Analysis, and so on), and (c) information that flows through the model. In general, uncertainty can be regarded as a breakdown in some or all of the parts in the planning model (inadequate tools, bad implementation of theory, irrelevant or outdated data).  In our case, we are mainly interested in how our external events, lack of knowledge of critical relationships, and change in the coefficients may invalidate the model and generate unwanted outcomes for the firm.

One way to incorporate uncertainty into our simultaneous equations model is to view it as an approximation.  The coefficients, equations, and relationships in our model are but a simplification of reality that we use as a positional reference system to support decision-making. Instead of working with the customary simultaneous equation system, A x = b, we may work with its approximate or interval form, AI x = bI .

**Exhibit 4.4**
**Main Relationships in the Financial Planning Decision**

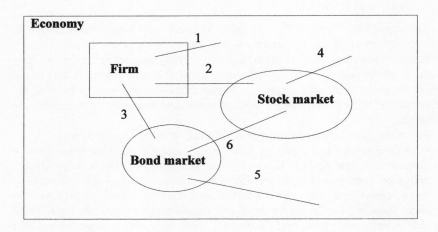

1. Sales and the economy
2. Equity financing
3. Fixed income financing
4. Stock market and the economy
5. Bond market and the economy
6. Stock market and bond market

## B. The Model, Its Specification and Solution

In sum, the resolution of the type of approximate equations generated by our planning problem starts by adding (and subtracting) an error of the same magnitude ($\varepsilon = \varepsilon ij$) from each coefficient in the original A x = b formulation. This error is carefully calculated to preserve invertibility of the coefficient matrix. In our case, from the numerical example of Exhibit 4.3, the maximum allowable error is 0.1336284. That is, if we add/subtract that figure to and from the original coefficients matrix in A x = b, w will be in trouble (noninvertible, noninsularity of the solution set, and so on).

For the example of reference this is good news, though, because it represents a very large percentage of some of the coefficients involved—for example, a12 = 0.034125, a13 = 0.001625. Note also that adding this error provides flexibility to other numbers in the model. For example, adding the full error to the target leverage ratio (0.5) means this ratio can fluctuate within the 86.63%–36.63% range. It is true we do not know the exact magnitude of error in our models but, in our example, it is reasonable to assume smaller values than the aforementioned maximum. We have calculated the effects of adding an error of 0.01 and 0.05 to the basic model, with the numbers from Exhibit 4.3. The case presented in this exhibit is interesting because it corresponds with self-sustained growth.

## C. Effects on Planning and Implications

Exhibit 4.5 presents the values for the approximate version of the simultaneous equations model. Our firm targets a self-sustained growth rate of 4.60%, which should result in retentions of 119.80 that would finance all growth needs. However, if we believe our model is approximate only up to 0.01 magnitude, retentions may fluctuate in the [121.1672, 118.46821] range, which means the firm may need external financing in the [2.5935697, -2.652163] range. These results would be consistent with a leverage ratio in the [0.51–0.49] range.

**Exhibit 4.5**
**Approximate Equations and Uncertainty**

Growth  0.046077935       (4.60%)

| Error | 0 | 0.01 | | 0.05 | |
|-------|----------|----------|----------|----------|----------|
|       | Centered | Max | Min | Max | Min |
| Rnew  | 119.8026 | 121.1672 | 118.46821 | 129.06595 | 111.77106 |
| Dnew  | 0.0000   | 2.1236681 | -2.1716455 | 12.781835 | -14.742094 |
| Snew  | 0.0000   | 3.369531 | -3.4424006 | 20.233952 | -23.180539 |
| FN    | 0.0000   | 2.5935697 | -2.652163 | 15.610056 | -18.004059 |

The last two columns in Exhibit 4.5 provide the corresponding results if the assumed approximation error is 0.05. With a 0.01 error, the planning mistake in earnings and external financing needs is only about 2.20%—for example, 2.20% ≈ 2.5935697/121.1672. With an approximation of 0.05, the planning mistake becomes a great deal more serious—for example, 12.10% ≈15.610056/129.06595.

Still, a useful, practical financial planning recommendation for small and medium-sized firms should be prepared for some surprises but stick to the plan of strengthening the self-sustained growth rate.

Note that the application itself, that is, adding errors to a fundamental model, took the least amount of effort in our long search for a practical model; but for us the theoretical content of the model is more important than the technical feature of adding errors. Before adding errors, one must know why, to what specific model, and on what grounds, otherwise we are dealing merely with technique.

In sum, implementation of an AES financial planning model follows these steps:

1.  Gather financial data concerning balance and income statements.
2.  Simplify them as much as desirable. Our knowledge of the firm is larger than that of the economy, still, it is wise to keep the model restricted to a few equations given its nature as a planning tool.
3.  Calculate the maximum error allowed and obtain ranges of uncertainty for the variables involved.
4.  Study past financial data to map observed data to those given by the model, so that the later becomes plausible and realistic.

This is, after all, what many management consultants and corporate financial planners do. In this sense, we are simply formalizing and refining practical financial planning. The particular suitability of our model for small firms adds a bonus to our research, since those firms are pervasive in every economy.

## SUMMARY

This chapter developed a practical financial planning model. The support obtained from economic theory was not very strong. From financial management, we reviewed the existent split between managing by parts— conventional financial management—and financial planning itself, which is an assortment of tools, techniques, methods, and two modeling approaches proper: simultaneous equations and linear programming.

The second section of this chapter concentrated on simultaneous equations and focused on the fundamental core of financial planning: assessing financing needs. It showed that a simple model with four equations serves to calculate financing needs and incorporates or is equivalent to other programming tools and methods.

The third and last section of this chapter reformulated the financial planning problem in approximate terms, which turned the fundamental planning model into a practical one.

In sum, a practical financial planning model particularly well-suited for small to medium-sized firms was presented. This model offers the following advantages over previous ones:

1.  Treats the firm as a planner.
2.  Incorporates financial planning methods and techniques.
3.  Stresses importance of self-reliance on internally financed growth, which is a prerequisite to any normative valuation model of the firm. Note that the model shows financing needs in dollar amounts, that is realistic and allows learning about capital and credit rationing that the firm may be facing.
4.  Permits assessment and optimal responses to changing economic environments.
5.  Is simple enough to both encapsulate major factors in financial planning and incorporate uncertainty, which are prerequisites for any model fit for practice.

Our simple model gives a straightforward but important prescription for firms—particularly small ones—to follow: strengthen the no-external-financing sustainable growth rate. It also provides specific optimal responses to changing conditions. These optimal responses require integration of financial and other management policies.

Viewing our model as an approximation is one way to incorporate uncertainty into the model and provide numerical and temporal relevance as well. The use of intervals validates the model in cognitive terms, as we seem to be more comfortable expressing expectations and policies in terms of intervals.

Viewing the financial planning model as an approximation adds the flexibility that practitioners need in academic models.

# 5

# PRACTICAL-APPROXIMATE-PORTFOLIO SELECTION MODELS

This chapter presents the last of our practical applications of approximate equations. The focus is on portfolio management, which covers the areas of security selection (determination of optimal weights), monitoring of optimal positions, and performance evaluation. The first of these areas, however, takes the lion's share in textbooks, which often allocate a few often foggy remarks concerning the customary distinction between "active" and "passive" management to the second area. As noted later on, portfolio theory is a single-period, static theory and, as such, has little to say about management of a given portfolio, which is a dynamic issue. The third area, performance evaluation, rests squarely on the contributions of the two main branches in portfolio analysis. One of them is called mean-variance analysis because it assumes the investment decision can be solely based on those two parameters. The other branch focuses on using indices in the investment decision and the better known of these models is the single-index model.

This chapter will concentrate on equity investing, even though the mean-variance approach can be applied to other securities as well.

It is important to stress that this chapter deals with many interlaced areas at the object level (statistics, linear algebra, mathematical programming), at the theory level, and the level of the individual securities and the specific markets to which portfolio analysis is applied. Moreover, each of these areas is experiencing profound changes simultaneously. This means that the chapter must compose a collage of information that may seem to have different levels of depth and breadth. It tries to provide as much background and specialized material—for example, that concerning the interplay between the risk free rate and short sales assumptions, quadratic programming and portfolio optimizations, and so on—as our discussion would require. Our overriding strategy has been to strengthen investment practice.

We have made decisions regarding what not to cover. For example, in investment analysis it is customary to start with thoughts on the economic environment. However, Chapter 3 faced some of the challenges posited by the macroeconomic components of the investment decision, and now it is merely assumed that the environment is favorable to equity investing. My decision is to dispense with utility theory completely. It never seemed anything else than a magician's trick to cover a hole in the fabric of the theories: sheer ignorance and lack of interest in what motivates human beings. What utility theory does is to manufacture an impressive appendage that is used to make theories look complete. Utility theory is not a door into learning about us, but a device to validate theories; the proof is that when an author uses utility theory, he or she becomes stagnant in the range of behaviors he or she will consider. We do not use utility theory—to prevent our applications from being overtaken by the type of permanent paralysis afflicting consumer theory—and note that this decision does not affect the practical import of the material included. The theorists, however, need not worry: A number of separation theorems—Fisher's, Tobin's, and Black's—indicate that the investment decision may be separated from individual utility considerations.

The evidence and testing of alternative equilibrium models are not covered either because the literature is immense, and ever-changing. In addition, there are already excellent surveys that have been used earlier, and such review is not necessary for this study. More importantly, I try to encourage and facilitate implementation in portfolio analysis. The following question has accompanied me from my graduate school days: How can you test if portfolio analysis creates efficient markets if you do not know whether investors actually use it?

The presentation of portfolio analysis concentrates on simultaneous equations systems (SES) and shows that it is possible to unify what appear to be "odds and ends," and even "loose ends," when this approach is used.

One important aspect of portfolio theory I want to emphasize is this: "Portfolio theory is a single period theory. Investors care only about the return on their portfolios over the interval, which begins when the portfolios are initially allocated and ends when the planning horizon ends. Investors are assumed to be indifferent to fluctuations in the values of their portfolio prior to the end of the planning horizon and indifferent to asset values following the end of the planning horizon." (Garbade, 1982, p. 107). In practice, of course, the liquidation dates among investors vary widely. One can think, however, that we will deal with periodic (yearly) revision dates. This makes the analysis more realistic than the more drastic concept of liquidation dates. In any case, the single period nature of portfolio theory strengthens the case for interval expressions of optimal weights and expected risk and returns, because some tolerance is needed if one is to maintain positions for extended periods of time.

The application of approximate equations to portfolio management is organized in a manner similar to previous applications in this book. First, it briefly presents the problem of security selection, the agents, and the context in which investment decisions are made. Second, it studies portfolio theory to find some suitable models. Finally, it obtains approximate portfolio weights and draws some preliminary conclusions. A summary and appendix concerning the portfolio literature will close the chapter.

## 1. INVESTMENTS, AGENTS, AND SECURITIES

We are all investors; everything we do is an investment decision of one kind or another. In the economic sphere of experience, investing concerns the administration of resources over time. "Over time" means we are forward-looking individuals, and that we are first and foremost planners. Resources include real and financial ones; as individuals, we are after real resources (homes, cars, college degrees, and so on), but we use financial ones—stocks, bonds, and so on—as vehicles to transfer over time the funds that will purchase those real assets.

There are many types of financial assets and they can be classified in a number of ways. One of the most useful distinctions is to break financial assets into two major classes: fixed income, representing debt, and variable income, representing equity. These two classes are the major pillars upon which the financial system is built, because they also represent two major ways by which firms raise funds, by selling bonds, equity, or both. The "asset allocation" and "leverage" problems of investors and firms refer to acquiring/issuing optimal proportions of these two major classes of assets given the operational and strategic objectives of each agent.

Another useful classification of financial assets is that between primary and derivative securities. One way to explain the difference is to say that primary securities represent claims on the issuing firm's assets, while derivative securities represent claims on the investor who trades those primary claims. For example, a company issues equity shares that are purchased by an agent who writes options on them—that is, writes calls to someone else. The option buyer has a claim on the trader, but not the firm that issued the stock. It is important to realize that derivatives trading was developed to create an activity that would enhance investors' opportunities to hedge and, thus, enhance returns and help mold expected cash flows. Better risk management for investors means they can optimize their trading and, thus, indirectly benefit firms by making sure their issues of debt or equity will be properly assessed and optimally traded. Most derivatives are issued by investors, not firms. What about warrants? Warrants are securities issued by firms to guarantee a purchase price for the company's stock or bonds and are usually issued together with new equity or bonds. In spite of the complexity and richness of activities in modern financial markets, there is still a very fundamental division that calls for firms to worry about producing goods and services and investors for trading firms' securities. Firms that invest in their own securities walk on the thinnest ice. First, they distract themselves from their primary activity; second, they may enter into dangerous legal and ethical grounds.

There are many types of derivative securities. The main types of contracts are forward, which are non-marketable, and futures, options, and options on futures, all of which are marketable. Options, futures, and options on futures trade in standardized blocks, which make it convenient for investors to trade them. With exception of the foreign exchange market, where forward contracts and banks play the major role, most of the growth in derivatives has come from options and futures, especially the so-called financial options and futures. These include options on common stocks, with shorter and longer maturities, options, and futures on stock market indices and treasuries packs. Large and specialized investors use those packs to take positions against their own portfolios of primary securities and, thus, hedge their original holdings.    Growth in derivatives has also extended to contracts on metals, agricultural goods, energy, cattle, foreign exchange, and so on.

There are also important sets of securities backed by real assets, especially of the fixed income kind, and financial assets that include derivative-like features—for example, convertible and callable debt.    There are also financial intermediaries that "securitize" cash flows; that is, they issue securities representing claims in pieces (or in groupings) of other securities' cash flows (for example, stripping coupon payments on bonds, or interest and principal payments in mortgage loans).    Finally, there are financial intermediaries that "manufacture" securities with specific characteristics in order to address complex needs of some agents (firms and other investors).    This is part of the process known as financial engineering.

There is a general way to reason about investments and investing that transcends the specific security discussed here.    Learning about investments and investing, in this context, is not so much learning the details concerning specific securities but the fundamental pricing-trading principles driving investors, companies, and markets.    This may explain why it may be more important the way one conceptualizes the investment decision itself, the decision-maker and its objectives, than whether our entry point into the world of investments is fixed income, derivatives, or equity.

There are different agents in the investments arena which, for purposes here, will be divided into institutional and individual ones.    This distinction is very important because it generates a great divide concerning investment techniques that are actually available to each investor.

Individual investors are also called small investors.    Most of them live on slow-changing wages, take care of a household, face significant transaction costs when they trade securities, and invest small sums of money at a time.    Asset allocation is very important for individual investors and includes decisions concerning real and financial assets.    Institutional investors, however, are specialized investors—investment companies, insurance companies, custodial or trust banks, pension funds, and so on.    This means they trade a particular set of securities (municipal bonds, or mortgage-backed securities), or a particular asset class—for example, fixed income.    Transaction costs are trivial for them because they trade in large volumes.    What Brennan said many years ago still applies: "(I)t seems reasonable to assume that the market is dominated by large investors for whom the fixed costs of transactions are trivial" (1975, p. 485). (Trading large volumes diminishes transaction costs but also makes it impossible for these investors to alter positions without "telling" other investors, up to the point that sometimes they have to build positions on particular companies or groups of stocks over a period of almost a year.)

In our capital markets we have different types of investors.    For example, Lease, Lewellen, and Schlarbaum found that: "Individuals do appear to partition themselves into distinct groups in terms of investment strategies, objectives, information sources, and asset selection behavior.    Furthermore, prior treatments of the segmentation issue—most of which have dealt with institutional investing—may, if anything, have underestimated its magnitude" Lease, Lewellen, and Schlarbaum (1976, p. 54). Their analysis of about 2,500 customers of a large brokerage house offered a view of individual (small) investors: two thirds of the customers were between 35 and 64 years old, 56% of the sample had an income of less than $25,000 (about $65,000 if we update that number to include inflation), 77% had at least a college degree, and they had an average number of eight securities in their portfolios.

**Exhibit 5.1**
**Life-Cycle Hypothesis**

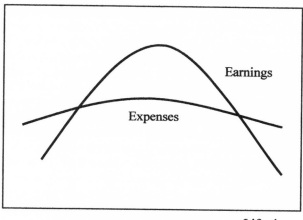

Market segmentation has presented problems for those developing investment theories. The usual response has been to standardize and homogenize individual investors into what we could call "Walrasian crowds." This modus operandi simplifies both the agent and the investment problem itself and, when in doubt, researchers have developed theories for larger investors, and also assumed small investors would use investment companies. The result is that we know little about, for example, the asset allocation problem, or the number of securities needed to optimize a portfolio. Brennan (1975) found that fixed transaction costs substantially reduce the optimal number of securities in a portfolio, especially in the case of moderate wealth levels, and suggested that it would no longer be necessarily optimal for small investors (reduced budgets) to hold perfectly diversified (efficient) portfolios in the presence of fixed transaction costs. Brennan's analysis shows that small investors should select a minimum level of risk and optimize the number of securities to hold in terms of expected returns.

We have only suspicions about how many stocks make a diversified portfolio. Optimization exercises found in textbooks often recommend at least 15 securities. Statman (1987) suggests about 40 and, more recently, Newbould and Poon (1993) double that estimate. Some of the issues that make this calculation difficult are the following: transaction costs, assumptions concerning short sales and the presence of a risk-free security, rate of return on the market portfolio, and the preliminary state of portfolio theory itself.

A great deal about investing can be learned when one faces difficult questions such as those facing individual investors. Take, for example, the problem of life-cycle investing. Exhibit 5.1 presents a plausible hypothesis concerning the needs of investors throughout their lives. At first, we need support until we obtain our first "real" job. Sometime after that we start earning more than we (need to) spend. These are the funds we use to support those who depend upon us, and that we will use during those years when our capacity (or our drive) to work long hours each day diminishes and we ultimately retire. Different societal arrangements—cultural or political—may modify this picture. For example, existence of public retirement systems may change the propensity to save during "high income" years. Still, in a country like the United States, the life-cycle picture tells that perhaps the proportions of fixed to variable income would change over the life of the individual. During the high-income years, investors should purchase equity because the money is not needed right away and it needs a long time to pay off. As retirement approaches, the proportion of wealth invested in fixed income versus variable income increases.

Compare the biological-life arrangements reasoning of the previous paragraph with what is called "cyclical investing," in which portfolio managers shift their fixed/variable allocations based on macroeconomic expectations. Students usually think most portfolio managers do that, but it is so difficult to anticipate economic cycles that only small proportions of portfolio managers (those of balanced funds) invest that way.

The life-cycle hypothesis also highlights that individual investors may not be able to invest the "one-shot" way that (single-period) portfolio theory implies. It seems for many individual small investors that the investment decision is a sequential one, and may require more qualitative than quantitative reasoning (see Tarrazo 1999, 1997a, and Smith 1974).

After reviewing some of the characteristics of investors, we must outline investment theories and how they were developed. The story starts with what we call the "time value of money," which is a set of formulas to value expected amounts at different points in time. Obviously, these formulas apply better to fixed than variable income, which varies in a manner difficult to anticipate precisely. This is what we call risk and the investment decision is about risk because it concerns itself with an uncertain future.

During the 1950s, Harry Markowitz offered a framework in which individual investors made their decisions in terms of statistical averages and variances, which would serve as indicators of expected returns and risk, respectively. Other researchers studied the implications of having Markowitz-type investors and formulated useful hypotheses about capital markets. The 1950s and 1960s represented the birth of portfolio theory, which was to be implemented (in a rather loose way) during the 1970s, especially through the explosive growth of investment companies. For instance, in 1975 there were 426 mutual funds (36 money market, 390 stock, bond and income), and in 1989 there were 2,918 mutual funds (664 money market, 2,254 stock, bond and income), based on data obtained from the Investment Company Institute. In 1997, there were over 6,000 mutual funds. More telling than the number is the value they manage, which is calculated at over 60% of the value of U.S. equities, see Radcliffe (1997, Chapter 14).

As noted, investing is planning, and planning has much to do with risk management. During the 1970s, we saw the formulation and application of what could be called the first level of risk management: diversification and duration-based immunization in fixed income portfolios. When interest rates fluctuate actual and expected reinvestment rates and capital gains of fixed income securities fluctuate as well, which makes actual (e.g., realized compound yields) and expected yields (e.g., yield-to-maturity) different. Duration-based immunization is a technique to calculate a holding period at which both yields will coincide. But the 1970s also saw the advent of option theory, which brought risk management to a new level based on the concept of hedging. This concept is more complex than diversification, which itself may be a form of hedging. Hedging may involve combining equity positions with assets of different classes—cash, short sales, common stock options, index products, fixed income, and so on. According to the Securities Exchange Act of 1934, a short sale is "the sale of a security that you do not own or if you own it you do not intend to deliver it." Generally the short seller borrows the securities from their owner, who keeps all the ownership privileges (for example, dividends), receives 100% collateral of the market value of the short sale, and receives interest payments on such collateral from the short seller.

During the 1990s we are immersed in yet another level of risk management: custom-made hedging programs. These are sometimes set up with synthetic securities through a process known as financial engineering.

These developments, however, have not rendered portfolio analysis obsolete but, perhaps, have reinforced the usefulness of basic risk management of the type provided by diversification. For example, these developments may have made equity markets more mature, sophisticated, and efficient, which would play to the strengths of portfolio analysis.

A dangerous misconception concerning market performance is to regard the index as an indicator of average returns: that is not so. Otherwise, the index would not consistently outperform 70% of other equity portfolios. The index may rather indicate quality and perhaps the best risk-adjusted return opportunities available to investors ($rp - rf/\sigma p$). This suggests investing on the index itself with what is called a "passive" strategy. Exhibit 4 in Wagner and Banks (1992, p. 8) shows that during the 1980s, active money managers fell short of index returns by an average 1.58% per year, see also Bogle (1992,

1994). Wagner and Banks (1992) argue that it is difficult for money managers to earn extra returns because competition makes it hard to constantly come up with fruitful new investments, and because research and quantitative trading has exploited gains from diversification. Wagner and Banks (1992) also find that professional money management firms may not realize how large transaction costs can actually become. "(T)he evidence from performance universes suggests that active managers frequently do not overcome the transaction cost handicap" (p. 7).

Our next step is to study portfolio models.

## 2. PORTFOLIO MODELS

We will first review the two main frameworks of portfolio theory, after which we will focus on how alternative assumptions and the optimization itself help or hinder formulation of portfolio selection in approximate form. The final subsection contains the models we will reformulate in interval form.

### A. Mean-Variance and the Single-Index Model

Exhibit 5.2 contains a summary of mean-variance analysis using conventional notation. The analyst using mean-variance techniques to select equity securities would do the following:

1. Select a group of "k" securities.
2. Obtain their closing prices (pt).
3. Calculate (one-period) returns. This can be done using either of the following formulas: (a) $rt = (pt - pt\text{-}1)/pt\text{-}1$, or (b) $rt = \ln(pt/pt\text{-}1)$.
4. Obtain estimates for their average values, variances, and covariances.
5. Select those securities that minimize the variance of the portfolio for a given level of portfolio returns (rp), and making sure the optimal weights add up to one—"full investment" assumption.

Of course, the previous list leaves some questions unanswered, which students quickly formulate, and which generally receive the following answers:

1. How many securities do we use? It is recommended to include about 15-20 securities. As noted earlier, Statman (1987) recommends about 40, and Newbould and Poon (1993) about 80. The number depends on the trading strategy (short sales versus no short sales) and other assumptions concerning how the sample is to be selected.
2. Do we calculate returns with or without dividends? If the stochastic process for dividends has a regular pattern, then dividends can be omitted. If the pattern is irregular and dividend payments are important, neglecting dividends will bias the optimization.
3. Do we use discrete or continuous compounding formulas? Either one is fine.
4. Do we use arithmetic or geometric returns? Try arithmetic ones.

5.  How do we run the optimization? With Excel's Solver, or with the matrix operations in Excel.
6.  What about cash holdings?
7.  What about short sales?

The final two questions require more than one-sentence responses. Let us start with the last one. The type of short sales generally used in textbooks ("standard" short sales) imply the seller receives short sales proceeds in full and no margin is required on shorted accounts: in reality, this is not the case. Markowitz, in his original formulation of the portfolio problem, precluded short sales altogether and used utility theory to determine which portfolio would be optimal for investors with different preferences. There will be more about short sales later on.

Without a risk-free rate and with or without short sales, we expect the optimization to provide what is called the efficient frontier when different required portfolio returns are used. An example of the efficient frontier is given by curve EF1 in Exhibit 5.3. It has a minimum variance portfolio, rv-σv, whose returns must be above the return on the risk-free rate (rf) for the exercise to have any meaning.

The issue of the risk-free rate (cash) requires a response organized into two parts. First, we must review what happens when a risk-free rate is included in the optimization, and then study how to get what is called the "unlevered," or tangent, portfolio. The first part can be explained with the help of Exhibit 5.3, and we will leave the calculation of the "unlevered" portfolio for later.

Suppose we had two types of assets: risky ones, which we could group into portfolio "r," and a riskless asset. Combining those assets into a portfolio "p" would yield the following return and risk:

(1)  $rp = wf\, rf + wr\, rr$

(2)  $\sigma p = wr\, \sigma r$

where wf, and wr are the weights—wealth allocations—in the risk-free asset and the portfolio of risky assets; rf and rr are the rates of return for each asset; and σp and σr the standard deviations of the portfolio and risky assets, respectively.

Noting that wf + wr = 1, and that wr = σp/σr, we can write

(3)  $rp = (1 - wr)\, rf + rr\, (\sigma p/\sigma r) = rf + [(rr - rf)\, (\sigma p\, /\sigma r)]$

**Exhibit 5.2**
**Mean-Variance Analysis**

---

No risk-free rate, short-sales allowed

$$\text{Minimize} \quad \sigma p^2 = W' \, \Omega \, W$$

Subject to     a) $W' R = rp$

b) $\Sigma \, wi = 1$

where:   $W' = [\ w1 \ w2 \ w3 \ ... \ wk\ ], R' = [\ r1 \ r2 \ r3 \ ... \ rk\ ]$

$wi$ = wealth proportions invested in security "i," $i = 1, 2, ... k$.

$ri$ = historical returns on security "i," $i = 1, 2, ... k$.

$\Omega$ = Variance-covariance matrix of historical returns, with typical elements $\sigma ii$ and $\sigma ij$

$\sigma ii = \sigma i^2$ = variance security $i$, and $\sigma ij$ = covariance securities $i$ and $j$

Lagrangian = $[-1/2\ \sigma p^2] - \lambda 1\ [\Sigma\ wi\ ri - rp\ ] - \lambda 2\ [\ \Sigma\ wi - 1\ ]$

| σ11 | σ12 | σ13 | ... | σ1k | r1 | 1 | | w1 | | 0 | (*) |
| σ21 | σ22 | σ23 | ... | σ2k | r2 | 1 | | w2 | = | 0 | |
| ... | ... | ... | ... | ... | ... | ... | | ... | | ... | |
| σk1 | σk2 | σk3 | ... | σkk | rk | 1 | | wk | | 0 | |
| r1 | r2 | r3 | ... | rk | 0 | 0 | | λ1 | | Rp | |
| 1 | 1 | 1 | ... | 1 | 0 | 0 | | λ2 | | 1 | |

A                                        X  =  C

---

No risk-free rate, no short-sales allowed

First step. Calculate the short sales solution as above.
Second step. Remove from the problem those variables with negative weights.

| σ11 | σ12 | σ13 | ... | σ1k-s | r1 | 1 | | w1 | | 0 | |
| σ21 | σ22 | σ23 | ... | σ2k-s | r2 | 1 | | w2 | = | 0 | |
| ... | ... | ... | ... | ... | ... | ... | | ... | | ... | |
| σk1 | σk2 | σk3 | ... | σk-sk-s | rk-s | 1 | | wk-s | | 0 | |
| r1 | r2 | r3 | ... | rk-s | 0 | 0 | | λ1 | | Rp (**) | |
| 1 | 1 | 1 | ... | 1 | 0 | 0 | | λ2 | | 1 | |

(*)     After some simplification.
(**)    The required rate of return (Rp) may need to be adjusted because under No-SS one cannot get the unrealistic portfolio returns usually obtainable under SS.

**Exhibit 5.3**
**Efficient Frontier, with and without a Risk-Free Asset**

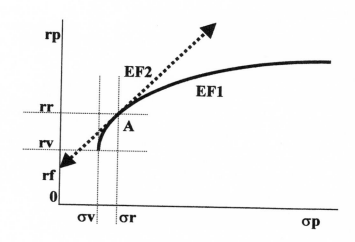

rp   = portfolio return

rr   = portfolio return on the unlevered (all risky assets) portfolio A

rv   = portfolio return on the minimum variance portfolio
       (note: rv > rf)

rf   = rate of return on the risk-free asset

σv  = standard deviation of the minimum variance portfolio

σr  = standard deviation of the unlevered (all risky assets) portfolio A

σp  = portfolio standard deviation

EF1 = efficient frontier without a risk-free asset

EF2 = efficient frontier with a risk-free asset.  The segment rf-A,
       represents lending portfolios, the segment above A represents
       borrowing portfolios.

This is the equation for the efficient frontier EF2 in Exhibit 5.3. Several important issues need comment. First, note that the efficient frontier is now a straight line, where a 100% cash position (wf = 1) would yield rp = rf, and a 100% position in risky assets (wf = 0, wr = 1) would yield rp = rr. Investing more than 100% in risky securities is possible, but will require borrowing at the rf rate. Investing less than 100% in risky securities implies we are lending. Second, note that portfolio "A" becomes the grouping of risky securities everyone would choose because it yields the best risk-adjusted premium—that is, the highest $[(rp - rf)/\sigma p]$. This is Tobin's (1958) "separation" theorem or, more properly, a "one fund argument." This argument—not to be confused with the two-fund theorem—says that any two optimal portfolios can span the entire efficient frontier, and implies that two optimal weights for a given security obtained from two efficient portfolios can generate all the optimal weight values for such security. At some point, the one-fund argument turns into the two-fund theorem; but this time with a money market and a single equity fund, which suffice to obtain all risk-return combinations along a linear efficient frontier tangent to the nonlinear one, and which passes through the points (rf, 0) and (rr, $\sigma r$). Third, note that an investor in this economy would have access to the entire return-risk spectrum by investing in two funds: a money market fund yielding rf, and an index fund—like portfolio A—yielding rr.

If we select a sample of "k" securities, we can easily calculate the tangent portfolio, with or without short sales, as we will show in Exhibit 5.7. That portfolio, however, could be uninteresting and maybe ill-built from an investments point of view as, for example, when it includes standard short sales. To fully appreciate what rf and rr represent, we must briefly review the theory of the capital asset pricing model developed by Sharpe, Lintner, and Mossin. These authors showed that in an economy populated by Markowitz-type investors, and under certain conditions, one could derive a pricing relationship (for securities as well as for portfolios) such as

(4)             $rp = rf + (rm - rf)\, betap$

This is the famous beta pricing relationship, where rm is the return on the market portfolio and betap is a coefficient that measures the sensitivity of our portfolio to the market portfolio. Beta can be estimated by regressing market returns on portfolio or individual security returns. The CAPM was originally derived under the following assumptions: (1) investors are of the Markowitz type, (2) there is a risk-free rate at which investors can borrow and lend, (3) investors have homogeneous expectations, (4) investments are infinitely divisible, (5) there are no taxes, transaction costs, or inflation, and (6) capital markets are in equilibrium. These are, of course, very restrictive assumptions but not all of them are needed to obtain the pricing relationship. The most important assumptions are that there is a risk-free rate and that the market portfolio is mean-variance efficient, see Ross (1977) and Elton and Gruber

(1995). Ross (1977) clarified why the efficiency of CAPM relationships depends on the efficiency of the market portfolio.)

Much of modern portfolio theory lies in equations (3) and (4). They are the same. In order to see this, we have to show (1) that $rr$ is actually $rm$, and (2) that beta is actually ($\sigma p / \sigma r$). With respect to the first question, note that the value of an index is that it indicates the best return-to-risk relationship. We can manufacture an index portfolio with our own sample, but we may miss what is most important for this index to indicate: quality in the stock market itself, not in an arbitrary group of securities. The market portfolio is not an average, it is a leader. In other words, portfolio A, when the overall market is evaluated, is the "alpha wolf" in the pack of managed funds. Combining any other portfolio along EF1 (no risk-free rate) with cash holdings would result in a portfolio with lower slope, $(rp - rf)/\sigma p$, than that of portfolio A.

With respect to the second question, the beta of a portfolio can be calculated by regressing market returns on the specific portfolio returns. The regression beta is $betap = \sigma pm/\sigma p\sigma m = \rho pm\ \sigma p/\sigma m$, where $\rho pm$ is the correlation coefficient between the portfolio and the market. By definition, efficient portfolios are those that are perfectly correlated with the market ($\rho pm = 1$), then $bp = \sigma p/\sigma m$, which ends our demonstration on the equivalence of practical equation (3) and theoretical equation (4). (Furthermore, when portfolio A is the market portfolio, its beta equals one and $\sigma p = \sigma m$, by definition.)

It was Sharpe who realized the important role that the concept of a market portfolio could play in investment management. He tackled this in his 1963 study resulting in his famous "diagonal model," which appears in Exhibit 5.4.

One could say the single-index model (SI) is simply a guilt-free version of the CAPM and, in many ways, it is simply that. What is relevant, however, is whether it enhances our investment decision-making, which it seems to do. Sharpe (1963) accomplished several objectives with his study:

1. Market returns, as approximated by the closing quotes of a widely used index, should be included in portfolio optimizations.
2. Alphas, as indicators of risk-premium ($\alpha i = ri - b\ rm$), should be included in the optimization.
3. Betas should be given a substantial role in the optimization because they summarize the relation of securities and portfolios to the mean and variance of the market: $bi = (\alpha i - ri)/rm$, $bi = \rho im\ \sigma i/\sigma m$. These formulae indicate that betas could be used as summary indicators of the quality of a portfolio or security.
4. The model is "simplified" in the sense that a full variance-covariance matrix is unnecessary. Note the sparseness of the matrix used in the optimization.

From a practical point of view, it seems to us that the value of the SI model lies in its use of the market index concept. Note, however, that the diagonal SI model does not use the risk-free rate, nor it is expected for a given grouping of securities to provide the market rate of return.

A great deal of effort has been put into finding a few sets of indicators of investment quality. However, it seems unrealistic to express such quality solely in a handful of variables. A great deal was made—at the time of Sharpe's article—of numerical issues concerning the calculations needed for mean-variance analysis.

**Exhibit 5.4**
**Sharpe's (1963) Single-Index "Diagonal" Model**

Minimize
$$i= k+1$$
$$\Sigma \quad wi^2 \; \sigma ei^2$$
$$i=1$$

Subject to

    a)
$$i= k+1$$
$$\Sigma \quad wi \; \alpha i = Rp$$
$$i=1$$

    b)
$$i=k+1$$
$$\Sigma \quad wi \; = 1$$
$$i=1$$

    c)
$$i=k$$
$$\Sigma \; wi \; \text{ßi} \; = wk+1$$
$$i=1$$

where

| | |
|---|---|
| $\alpha(k+1)$ | $= rm$, and $\sigma^2 e(k+1) = \sigma m^2$ |
| bi | $= \sigma im / \sigma i \sigma m$ = slope from regressing rm on ri |
| $\alpha i$ | = rmean value for ri − (ßi * mean value for rm) = intercept of regression of rm on ri. |
| $\sigma ei^2$ | $= \sigma i^2$ - ßi $\sigma m^2$ = residual variance from regressing rm on ri |
| $\sigma ii$ | $= \text{ß}i^2 \sigma m^2 + \sigma ei^2$ = variance or ri decomposed into market and non-market risk |
| ßp | $= \Sigma wi \; \text{ßi}$ = Beta of the portfolio |

or, in matrix form

| | | | | | | | | | | |
|---|---|---|---|---|---|---|---|---|---|---|
| $2\sigma e12$ | 0 | ... | 0 | $\alpha1$ | 1 | $\beta1$ | $w1$ | | 0 | |
| 0 | $2\sigma e22$ | ... | 0 | $\alpha2$ | 1 | $\beta2$ | $w2$ | | 0 | |
| ... | ... | ... | ... | .... | ... | ... | ... | | ... | |
| 0 | 0 | ... | $\sigma m2$ | $\alpha k+1$ | 1 | -1 | $wk+1$ | = | 0 | |
| $\alpha1$ | $\alpha2$ | ... | $\alpha k+1$ | 0 | 0 | 0 | $\lambda1$ | | Rp | |
| 1 | 1 | ... | 1 | 0 | 0 | 0 | $\lambda2$ | | 1 | |
| $\beta1$ | $\beta2$ | ... | -1 | 0 | 0 | 0 | $\lambda3$ | | 0 | |
| | | | A | | | | X | = | C | |

**Exhibit 5.4 (Continued)**

For a case of two risky securities:

$$
\begin{vmatrix}
2\sigma e12 & 0 & 0 & \alpha1 & 1 & \beta1 \\
0 & 2\sigma e22 & 0 & \alpha2 & 1 & \beta2 \\
0 & 0 & 2\sigma m2 & rm & 1 & -1 \\
\alpha1 & \alpha2 & rm & 0 & 0 & 0 \\
1 & 1 & 1 & 0 & 0 & 0 \\
\beta1 & \beta2 & -1 & 0 & 0 & 0
\end{vmatrix}
\begin{vmatrix}
w1 \\ w2 \\ wk+1 \\ \lambda1 \\ \lambda2 \\ \lambda3
\end{vmatrix}
=
\begin{vmatrix}
0 \\ 0 \\ 0 \\ Rp \\ 1 \\ 0
\end{vmatrix}
$$

Note:   a) This specification does use the mean value for rm, explicitly, since
$\alpha k+1 = \alpha m = (1/n)\Sigma\ rtm = rm$
b) A basis reduction method can be applied to this specification to obtain
the no-short sales case.

---

It seems to us that this never should have been an issue given the developments in raw computing power and, very especially, considering there are no studies that show one needs more than a hundred securities to build optimal portfolios. Furthermore, as noted later, quadratic programming is not needed for portfolio optimization, which can be implemented with SES methods even for the no short sales case, as is noted in Exhibits 5.2, 5.4, and 5.5.

Exhibit 5.5 presents a nondiagonal version of Sharpe's "diagonal model," which we have never seen in the bibliography. As Sharpe himself noted, the presence of zeroes speeds up computations, which seems to favor the diagonal model at the time the model was formulated. However, remember that we rarely need more than a hundred securities in an optimal portfolio, and that we now have ample computer power even in household computers, which can handle sophisticated optimizations with even a few hundred variables. Therefore, the choice between the diagonal model and its nondiagonal counterpart should be evaluated in a different manner. There are two important items to be stressed. First, the nondiagonal version shows the real closeness between mean-variance and single-index formulations, which differ only in exclusion of residual covariances in off-diagonal positions of the variance-covariance matrix. In a way, the SI closely resembles mean-variance analysis itself. That is precisely why we like it. Second, the presence of zeroes in a matrix may speed computations considerably—think about Laplace expansions—but it may also indicate fragility of relationships, or even holes in our knowledge of the system under analysis. Both of these possibilities may signal ill conditioning in the matrices under evaluation (Jacobians from first-order conditions).

Exhibits 5.4–5.5 also present some notes concerning the single index-EGP algorithm, which calculates the tangent portfolio A for a specific user-selected sample—that is when our sample may not yield the rm itself. EGP algorithms are methods to obtain tangent portfolios for the single-index and mean-variance (assuming constant correlation) cases.   It is particularly well suited for spreadsheet manipulation, can calculate both short and no-short sales cases, and is easy to implement, see Elton and Gruber (1995).   EGP algorithms always include the risk-free rate, since they optimize (rp-rf)/bp, or (rp-rf)/σp, for the single-index and constant-correlation cases, respectively.

At this point, we have met the objectives of reviewing the two main models of portfolio selection.  Our discussion has not exhausted all the topics of interest concerning the single-index model.  For example, what is the exact role played by αi's?  Shouldn't we improve our sampling "effort" until our rp equals rm? What is the role of the risk-free rate in the single-index model? We will come back to some of these questions (especially the last one) but now it is time to close this subsection by mentioning other portfolio selection models and briefly summarizing what we have seen up to now.

Many authors have tried to extend and supplement the single-index model and its supporting theory (the CAPM).   One such effort is the arbitrage pricing theory developed by Ross (1976), which relaxes the assumption of the CAPM and creates room for other factors to be included in the pricing equation, see also Roll and Ross (1984).  Practical models incorporating the principles of the APT may be, however, indistinguishable from multi-index models, see Elton and Gruber (1995).

Another major effort refers to multiasset—and multimarket—equilibrium pricing.  This approach stresses the linkage between primary securities and their derivatives, which may be instrumental to develop better models for risk structure of the economy.  Financial futures in both equity and fixed income markets are so prominent that a theory that does not include them may be hardly plausible.   Increasing integration and dependency of segments within each financial market, which makes it difficult to build partial models, also require a multiasset approach.   The promising work of integrating different financial theories is progressing,  see Connor (1984), Hsia (1981), and Cox, Ingersoll, and Ross (1985), but it may take some time to obtain results that may improve investment practice.

In the meantime, practitioners can sample betas built under a variety of objectives, such as taking care of industry effects, incorporating accounting data in the beta regression ("fundamental" betas), and modeling betas with particular stochastic specifications.

How good are portfolio models?  The answer to this question is sought at two levels: (1) at the investor's level, and (2) at the market level.  Unfortunately, there are few studies on how particular investors fare when they use portfolio analysis.   Most of the literature has tried to check whether the theoretical constructs of portfolio analysis fit reality, with the implication that they should be good practical models if they are shown to be good descriptors of financial markets.  This approach, while somewhat meritorious, falls short of providing solutions to practical problems and is not practical itself: it takes too long—

forever—and, meanwhile, investors make decisions on a daily basis and cannot wait until researchers find out how the economy and financial markets work.

**Exhibit 5.5**
**Single-Index Nondiagonal Model**

---

General single-index model, No risk-free rate, short sales allowed

Minimize

$$\sigma^2 p = \beta p^2 \; \sigma m^2 + \Sigma w i^2 \; \sigma e i^2 = W' \; \Omega W$$

Subject to

    a)   $\Sigma w i \; r i = r p = W' \; R$
    b)   $\Sigma w i \; = 1$

where   $b i = \sigma i m / \sigma i \sigma m$ = slope from regressing rm on ri
             $\alpha i$ = rmean value for ri – ($\beta i$ * mean value for rm)  (intercept
                of regression of rm on ri)
             $\sigma e i^2 = \sigma i^2 - \beta i \; \sigma m^2$   (residual variance from regressing rm on
                ri)
             $\beta p = \Sigma w i \; \beta i$ (Beta of the portfolio)
             W' = [ w1  w2  w3  ...  wk ], R is a (k x 1) vector of returns

Lagrangian = $[-1/2 \; \sigma p^2] - \lambda 1 \; [\Sigma \; w i \; r i - r p ] - \lambda 2 \; [ \Sigma \; w i - 1 ]$

$$
\begin{vmatrix}
\sigma 11 & \sigma 12 & \sigma 13 & ... & \sigma 1k & r1 & 1 \\
\sigma 21 & \sigma 22 & \sigma 23 & ... & \sigma 2k & r2 & 1 \\
... & ... & ... & ... & ... & ... & ... \\
\sigma k1 & \sigma k2 & \sigma k3 & ... & \sigma kk & rk & 1 \\
r1 & r2 & r3 & ... & rk & 0 & 0 \\
1 & 1 & 1 & ... & 1 & 0 & 0
\end{vmatrix}
\begin{vmatrix}
w1 \\ w2 \\ ... \\ wk \\ \lambda 1 \\ \lambda 2
\end{vmatrix}
=
\begin{vmatrix}
0 \\ 0 \\ ... \\ 0 \\ Rp \\ 1
\end{vmatrix}
$$

Or, noting that $\sigma i i = \beta i^2 \sigma m^2 + \sigma e i^2$ ; $\sigma i j = \beta i \beta j \sigma m^2$, for $i \neq j$ ; $r i = 1/n \; (\Sigma \; r t i) =$ average value for the series rti, and after some simplification.

$$
\begin{vmatrix}
(\beta 1^2 \sigma m^2 + \sigma e 1^2) & \beta 1 \beta 2 \sigma m^2 & \beta 1 \beta 3 \sigma m^2 & ... & \beta 1 \beta k \sigma m^2 & r1 & 1 \\
\beta 2 \beta 1 \sigma m^2 & (\beta 2^2 \sigma m^2 + \sigma e 2^2) & \beta 2 \beta 3 \sigma m^2 & ... & \beta 2 \beta k \sigma m^2 & r2 & 1 \\
... & ... & ... & ... & ... & ... & ... \\
\beta k \beta 1 \sigma m^2 & \beta k \beta 2 \sigma m^2 & \beta k \beta 3 \sigma m^2 & ... & (\beta k^2 \sigma m^2 + \sigma e k^2) & rk & 1 \\
r1 & r2 & r3 & ... & rk & 0 & 0 \\
1 & 1 & 1 & ... & 1 & 0 & 0
\end{vmatrix}
\begin{vmatrix}
w1 \\ w2 \\ ... \\ wk \\ \lambda 1 \\ \lambda 2
\end{vmatrix}
=
\begin{vmatrix}
0 \\ 0 \\ 0 \\ 0 \\ Rp \\ 1
\end{vmatrix}
$$

                  A                           X   =   C

Notes:    (a) This specification does not use the mean value for rm = (1/n)$\Sigma$ rtm.
            (b) This specification uses average returns for each security (ri) , while Sharpe's version uses $\alpha i$. Both of them are related since $\alpha i = r i - \beta i \; r m + e i$.

**Exhibit 5.5 (Continued)**

(c) The Elton-Gruber-Padberg single index algorithm (EGP-SI) uses average returns, ri, also.

(d) The EGP-SI uses risk premia (ri = ri –rf). In the simultaneous equations version, weights are unaffected by using risk premia (ri-rf) as long as rp is also adjusted to rp = rp –rf. Using risk premia does not mean we are "including" the risk-free rate in the sense of calculating a tangent portfolio, unless rp = rr.

(e) A basis reduction method can be applied to this specification to obtain the no-short sales case.

Another problem with the conventional method of testing the effectiveness of portfolio theory is that it drags other hypotheses into the analysis, some of which need testing themselves. These are not simple hypotheses, but grandiose and thoroughly hard to prove statements about the whole of financial markets, such as, for example, whether they are efficient. In this it is not hard to see another manifestation of that specter that plagues economists' work: a built-in, inherited inertia to prove the superiority of the market system via efficient prices. Evidence for such contention is that studies on utility end up fairly soon as discussions of Pareto optimality.

The lack of responsiveness of conventional research to practitioners' daily dilemmas and the problems mentioned above exasperate money managers and some researchers as well, as can be observed in Harrington 's (1987) study on using the CAPM and the APT, and in Haugen's recent research (1999a, 1999b, 1995), who seems to advise judicious use of market theories so long as they do not impair our judgment by becoming dogma. In a way, Haugen's approach regarding market efficiency, normative pricing, and other "family pets" of finance scholars is reminiscent of Russell's skepticism, exemplified by the following quote: "I wish to propose for the reader's favorable consideration a doctrine which may, I fear, appear wildly paradoxical and subversive. The doctrine in question is this: that it is undesirable to believe a proposition when there is no ground whatever for supposing it true" (From Ayer, 1972, p. 18).

The reader interested in reviewing the current evidence on market efficiency/portfolio theory is referred to the excellent surveys in textbooks such as Elton and Gruber (1995, Chapter 17), and Haugen (1997, Chapters 8, 23, and 24).

The following observations have been drawn from the literature and my own experience and are relevant for this research on approximate portfolio analysis:

1. Descriptive statistics such as mean, variance, and covariance may carry useful information about securities and about the cross-sectional (static) risk-structure of the economy.

2. Mean-variance analysis provides a unique framework for taking advantage of statistical information. It may not incorporate all the information that is relevant for investment decisions but it may have some value.

3. Estimates for the risk-free asset and the market add valuable information to support the investment decision. Beta cannot be used as the sole indicator of risk and return; it is a worse indicator of return than it is of risk. In all, we believe the value of the single-index model resides in the overall model—risk-free rate and market estimates—and not in particular elements of the model (for example, betas, and alphas).

4. Market index estimates tend to eclipse the effect of other potential variables in the determination of stock returns.

5. The distinction between systematic and nonsystematic risk may be inaccurate but it is useful in thinking about economy-wide and firm-specific factors.

6. Portfolio analysis provides the only comprehensive framework for organizing investment reasoning, since it contains not only pieces concerning security selection, monitoring, and performance, but also how to integrate them. There is simply nothing comparable in investment analysis.

7. One cannot enhance the management of a group of stocks with index derivatives without knowing a great deal about the risk and return of such a portfolio. One needs measurements of risk and return in order to set up hedges, which would be very costly or unreliable if such measurements were not available. This is an important shortcoming of fundamental analysis that, because of it, is likely to provide portfolios plagued by common factors (poorly diversified, companies sharing the same characteristics), and that do not lend themselves to hedging via index products. Cash hedging, of course, is always possible but other types of portfolio hedging (using bonds or other assets) are difficult to implement because of the lack of exposure measures. Such "fundamental" portfolios can also be hedged with common stock options, see Sears and Trennepohl (1993, Chapter 19) and Bookstaber and Clarke (1981) and references therein, but this is an inferior method to hedging with index products because, in general, it is harder and more laborious to estimate hedge ratios at the security level than at the portfolio level itself. This observation is made in the context of large portfolios. Small (non-diversified) portfolios, however, may benefit from both the increased knowledge of securities that fundamental analysis may provide and stronger links between each security and its derivative (see Tarrazo, 1990).

At the current state of knowledge, we would expect all investors to make use of basic notions in portfolio theory such as diversification, and at least institutional investors to calculate optimal portfolio weights as another piece of information to guide and support their trading decisions. However, we would not recommend mechanical implementation of portfolio analysis to support automatic trading.

## B. From Theory to Practice

In academia, we sometimes tell students in investments courses that investment managers seem to have three choices regarding how to manage an equity portfolio: (1) technical analysis, (2) fundamental analysis, and (3) portfolio analysis. Technical analysis is difficult to learn; we also feel it resembles Sumerian astrology, and never understood its cabalistic procedures and incantations. Moreover, we would never like to explain to a panel of jurors how, even though we lost money, we selected our "winners" with fancy charts, and Japanese candlesticks (and quite possibly tea leaves). Therefore, we always cut the choices to two: (1) fundamental analysis, and (2) portfolio management. Fundamental analysis is appealing because it thoroughly studies the economy, the market, industries, and companies. It also appears preferable when we invest in a few stocks or particular industries, or for trading companies that go through the type of special situations that would render statistical data irrelevant (for example, mergers and acquisitions, bankruptcy, awards or losses of government contracts, unusual growth patterns, lawsuits, and so on). Fundamental analysis, however, is ill-prepared to scientifically manage large portfolios because it provides no indication of return-risk at the portfolio level. As noted earlier, without these indicators one can hardly take advantage of advanced hedging techniques. For these reasons, one would think that even portfolios that are managed along fundamental lines could benefit from using portfolio indicators.

What are the obstacles to using portfolio analysis? There are several obstacles, many of which have been clearly outlined by Michaud (1989, 1998):

1.  *Conceptual demands on portfolio managers.* Michaud mentions that many portfolio managers are used to other concepts that reflect a more "informal" tradition of investment management.
2.  *Political issues.* Introduction of optimizers shifts the political power toward "quants," which traditional managers dislike.
3.  *Practical.* Portfolios obtained from optimizers are sometimes not intuitive, without obvious investment value, and often difficult to market.
4.  *Limitations of the model itself.* First, the issue of error maximization; Michaud and others (for example, Kritzman, 1992) utilize Jobson and Korkie's (1980, 1981) results to argue that mean-variance portfolios are unreliable because the optimizations are not clean from residuals and randomness. Michaud also mentions problems in estimating the parameters, mixing factors in mean-variance approximations, and nonunique and unstable optimal solutions.

Two of these obstacles need some comment. First, the "practical" obstacle may indicate that analysts took a rather large sample of securities, applied portfolio optimization, and obtained a hard-to-understand combination of little known stocks that were, therefore, difficult to market. We believe it is better to apply portfolio selection to further improve a group of selected stocks. In this way, one may reap the benefits of portfolio theory while preserving some of the "easy to market" stocks. Unfortunately, nothing in portfolio theory suggests guidelines for construction of such a sample. To the contrary, one may think that choosing some stocks on the basis of shared attributes might result in returns bias and higher portfolio variance. It is even more dramatic to observe that one can choose 10–15 securities at random and calculate portfolio returns and standard deviation under equal weights to obtain market returns at market risk. Some may take this fact as overwhelming evidence of market efficiency.

The "error maximization" argument against mean-variance theory needs some comment. Jobson and Korkie (1980, 1981) are a textbook case of spurious results.  First, because they allow for unrestricted (or standard) short sales in their optimizations.  Second they count as "optimal" those portfolios of risky securities that yield returns below the risk-free rate!  This means their investors would bear risk and obtain returns lower than the risk-free rate (rather than simply taking their money to the bank).  Jobson and Korkie acknowledge this fact in their study (1981, p. 544 concerning short sales, and p. 553 regarding risky returns below the risk-free rate; see also Merton, 1972, and Frost and Savarino 1988).  In both theory and practice, risky portfolios with returns below the risk-free rate make no economic sense.  In practice, again, no one can use portfolios with unrestricted short sales.  However, even though it is clearly illogical to criticize the practical value of mean-variance analysis based on that type of evidence, some authors seem to be fond of doing so.  Alternative short sales assumptions also get in the way of assessing portfolio requirements as the requirement that the optimal weight of a security in a portfolio must not exceed 5% of the market value of the portfolio, see Tarrazo (1998b).

We believe some of the factors that hamper implementation of portfolio analysis, in addition to some of the factors already mentioned by Michaud (1989, 1998), are the assumptions employed in portfolio optimizations and the numerical optimizations themselves.  It is important for the purposes of this monograph that we study both of these factors, which we will do in the next subsection.  Formulating portfolio optimizations as approximations should address some of the issues concerning estimation, accuracy, and reliability of optimal weights and model selection risk.

## 1. The Role of Assumptions

There are many portfolio optimization scenarios.  The following list enumerates at least eight of them:

| Short sales: | Risk-free rate Not allowed | Allowed |
|---|---|---|
| Not allowed | 1 | 2 |
| Standard | 3 | 4 |
| Lintnerian | 5 | 6 |
| Realistic | 7 | 8 |

The most realistic among these cases would include the risk-free rate and avoid unrealistic short sales scenarios.  However, these cases have not been the most popular ones in textbooks because when the risk-free rate is included we use only a single portfolio of risky assets, which opens the door to many unanswered questions.  Further, calculating EF1 requires a little understood weight rebalancing technique, which will be reviewed next.  The risk-free rate brings in information that complements that of the market rate of return; it seems to act as an inspector that would reject securities and portfolios that do not meet minimum requirements.  Further, with the risk-free rate we also bring in cash management by borrowing and lending, and a hedging alternative to short sales.

There is no question that portfolio optimizations should include the risk-free rate to let the analyst assess investment opportunities represented by the line EF1.

As noted earlier, it is not true that one can calculate the unlevered, tangent portfolio when the risk-free rate is allowed by simply using risk-free-adjusted returns (ri' = ri –rf, and rp' = rp- rf) in models where a required return on the portfolio must be input (Exhibits 5.2, 5.4, and 5.5). In this specification, rp must be equal to rr. The problem is that rr (and rp) cannot be ascertained beforehand. What is usually done, in addition to using risk-free-adjusted returns, is to rebalance weights, so that wi* = wi'/Σ wi'. That is we rebalance weights so that they add up to one. Then we use these wi*'s to calculate the return on the tangent portfolio (rr). The best expositions of this rebalancing technique can be found in Levy and Sarnat (1984, two different ways, see p. 317 and ss.), and in Elton and Gruber (1995, p. 100). These expositions also involve excluding the Rp constraint in the first way described by Levy and Sarnat, and both Rp and the full wealth (Σwi = 1) or affinity constraint in Elton and Gruber's way, which is identical to the second way described by Levy and Sarnat. The weight rebalancing technique is the backdoor way to reinclude such constraints. The regular Elton-Gruber-Padberg algorithms calculate tangent portfolios because they incorporate the weight rebalancing technique. We can also use quadratic programming to maximize the Sharpe (ri-rf/σi), or Treynor (ri-rf)/βi performance index and, in these cases, calculate the tangent portfolio. However, quadratic programming can also be used to minimize portfolio variance subject to a given return, and subject to the full investment constraint without any inclusion of a risk-free rate.

The no short sales case may have been unpopular in textbooks because it is thought that its resolution requires quadratic programming, which until recently would require specialized software. As noted later, a no short sales optimization can be obtained from the short sales case by eliminating those variables with negative weights. Alternatives to the no short sales case are unappealing. Standard short sales, in which the investor receives the full proceeds from the short sale and faces no margin requirement, are unrealistic and illegal. Lintner's case, which assumes the proceeds are kept on deposit by the broker, implies a 100% margin on short sales positions and that interest is paid on both proceeds deposits and margin. This case is also unrealistic.

Lintnerian short sales can be calculated by requiring the sum of absolute weights to equal one, instead of the usual affinity constraint—sum of weights equals one. Investment on Lintnerian short sales can be regarded as one in which the agent purchases bonds yielding the risk-free rate (Levy and Sarnat, 1984) in conjunction with the purchase and short selling of stocks. Lintner's short sales have no meaning without a risk-free rate (see p. 32 in Elton and Gruber, 1995). For example, if an investor sells short $10,000 under Lintnerian short sales, the investor receives no cash from the transaction and, on top of it, is required to put up another $10,000 in a margin account (100% margin). Without a risk-free rate, however, the investor is foregoing too much money on the transaction—if rf = 5%, she would forego $1,000 with certainty, that is 5% on $20,000, or 10% on the $10,000.

Assuming a margin of zero, or one, is uninteresting from a practical viewpoint, since in many countries short sales are subject to other margin requirements.  In the United States, for example, investors are required to comply with Regulation T of the Federal Reserve, and exchanges may impose additional rules especially concerning minimum maintenance requirements. For example Rule 431 for maintenance margins in the NYSE.  The margin rules of the New York, regional exchanges, and the NASD are identical.  Curley (1989) is a useful reference for learning about specific details concerning trading and margin requirements.

In a realistic short sales case, the investor is required to keep a margin of 50%—Regulation T, initial margin requirement—and will receive interest payments on margined accounts (rf1), which are customarily paid, and also on the short sales deposits (rf2, rf2 < rf1).  The later payments depend on the agreement between the investor and the broker.  Margin requirements and the payment of interest modify the economic and accounting procedures of the realistic short sales portfolio problem.  Alexander (1993, 1995), who proposed a method based on the Elton-Gruber-Padberg algorithm, made a breakthrough in this area.   Tarrazo and Alves (1998b) clarified Alexander's numerical procedures and proposed a mixed-integer model especially suited to realistic short sales cases.

The issue of the risk-free rate rolls back into the short sales discussion and justifiably so: first, because interest payments on margin and short sales proceeds are important to the short seller; second, because cash is a powerful alternative to short selling.   Understanding the implications of these two observations may bring portfolio theory closer to practitioners than any other development.

We have seen that textbook cases that exclude risk-free asset and no short sales restrictions might make it difficult for practitioners to "buy into" portfolio theory.  The usual treatments of the optimization itself are also likely to drive practitioners away.

## 2. The Optimization Itself

Exhibit 5.6 lists the numerical procedures that can be applied to solve a portfolio under alternative restrictions concerning short sales and the risk-free asset.  The portfolio optimization problem can be solved always with quadratic programming (QP), in some cases with the Elton-Gruber-Padberg algorithm (EGP), and always with SES techniques, which focus on the first order conditions necessary for the optimization.

QP can be used to solve any portfolio optimization but obscures the economic rationale—that is, security selection—behind numerical procedures.   This monograph and Tarrazo (2000b) indicate why QP is unnecessary to optimize a portfolio. Let us review the EGP method before outlining Tarrazo's (2000b) assertion.

The EGP algorithm, mentioned earlier, has few shortcomings and teaches much to the discerning observer. It calculates a tangent, or unlevered, portfolio using the SI model that can be used to calculate the efficient frontiers EF1—by varying wf—and EF2—by varying the risk-free rate. In the mean-variance case, it uses a constant correlation assumption to calculate a tangent portfolio. Without that assumption, they could not calculate a parsimonious rule to derive the portfolio variance.

The EGP algorithm teaches that what makes a security attractive is its risk-adjusted premium—$(r_i-r_f/\sigma_i)$ or $(r_i-r_f)/\beta_i$, which also determines the security ranking within a portfolio. It also shows that the ranking of risk-adjusted premia is exactly identical to the ranking of portfolio weights. Covariance plays no role in the ranking of securities. This can be explained by noting that the variance-covariance matrix in any portfolio optimization is diagonally dominant, which means that elements on the main diagonal have more impact in the solution than off-diagonal elements. With the EGP it is easily observed that all securities in the sample are either bought or sold short (standard, Lintner, and realistic cases). Under no short sales, the cut-off point may exclude securities from the sample. In some cases our initial sample of securities may not require that any security be sold short. Finally, the EGP method lends itself well to spreadsheet manipulation and leads to further analysis of portfolio issues, as in Alexander (1995). On the minus side, however, it does not stress the role of rm as much as we would like. The market variance is explicitly used in the EGP algorithm, but the market returns are nowhere to be seen. This pinpoints an ambiguity of the single-index model in that we use rm for some things but not others.

Now we can review the simultaneous equations (SES) or linear algebra method. Exhibits 5.2, 5.4, 5.5, and 5.6 noted that a basis reduction method could be applied to the short sales solution to obtain the no short sales case by removing from the optimization those variables with negative weights. This was first noted by Martin back in 1955, and touched upon by Francis and Archer (1971), and again by Francis (1986, especially p. 769). However, none of these authors fully explored the idea, perhaps because it was unwise at the time to show that quadratic programming was not essential for portfolio optimization. We must remember that the period from 1950 to 1980 was when mathematical programming exploded into almost every field of quantitative research. Tarrazo (2000b) notes first that it is well known that Lagrangian optimizations can solve quadratic programming problems subject to linear constraints by substituting those constraints with special variables from then onward called Lagrangians (see Theil and Van de Panne, 1960, and Boot, 1964). In this case, first order conditions yield a simple linear simultaneous equation system. Linearity is not the issue; the problem stems from non-negativity constraints. The remedy is simple, eliminate variables that need negative weights from the optimization and one obtains an all-positive solution. Programming methods work by using slack variables for equations (for example, $w_i \geq 0$) that are binding in the optimal solution. If all the variables have positive values in the final solution, one can dispense with non-negativity constraints because they would be trivial (nonbinding) anyway.

**Exhibit 5.6**
**Portfolio Optimizations: Assumptions and Numerical Procedures**

|  | Mean-variance | Single-index |
|---|---|---|
| **Short sales:** |  |  |
| 1. Not allowed | QP, EGP*, SES* | QP, EGP, SES* |
| 2.1 Standard | QP, EGP*, SES | QP, EGP, SES |
| 2.2 Lintnerian | QP, EGP*, SES | QP, EGP, SES |
| 2.3 Realistic | QP, EGP*, SES | QP, EGP, SES |
| **Risk-free rate:** |  |  |
| 1. Not allowed | QP, SES | QP, EGP, SES |
| 2. Allowed | QP, EGP*, SES** | QP, EGP, SES** |

|  | Risk-free rate Excluded | Risk-free rate Included |
|---|---|---|
| **Short sales** |  |  |
| Excluded | SES | SES** |
| Included | SES* | SES* |

| QP | = Quadratic programming |
|---|---|
| EGP | = Elton-Gruber-Padberg algorithm |
| EGP* | = Elton-Gruber-Padberg constant correlation algorithm |
| SES | = Simple simultaneous equations solution from Lagrangean optimization |
| SES* | = Two step SES optimization to implement a basis reduction method |
| SES** | = Iterative basis reduction method |

Tarrazo (2000b) shows that basis reduction makes sense in both economic and mathematical terms, and also proofs why. Economically, one can see with the help of the EGP algorithm that variables with negative weights (securities to be sold short) are excluded from the no short sales case because they fail to score above a given cut-off value. It is not economically feasible to purchase them because their risk-adjusted premium is insufficient. In mathematical terms, Tarrazo (2000b) shows that Farkas's theorem (duality in programming), the Fredholm alternative of linear algebra, and both Farkas's and Khun-Tucker theorems would not hold if the basis reduction method did not work. It is rarely mentioned that non-negativity restrictions are rather special and dealt with in a very particular manner in mathematical programming, since their exclusion is implemented via an exclusion rule—see Foulds (1981, p. 359 and ss.) of Wolfe's (1959) algorithm, or Wolfe's (1959) original article.   The basis reduction applies the very same exclusion rule in a two-step, or multi-step in some cases, fashion.

One of the objections raised against the practicality of portfolio analysis is that mathematical programming makes it hard to implement or understand.   By showing that portfolio optimizations do not need quadratic programming, we demolish that objection.   We have also found that dissecting the portfolio optimization problem is useful to learn how economic rationale drives numerical procedures.

It is not uncommon to find the following observation in investment textbooks: "In the financial industry there are multitudes of special purpose programs designed to solve this problem (quadratic programming-based portfolio optimization) for hundreds or even thousands of assets" Luenberger (1997, p. 161). This seems to indicate that implementation of portfolio analysis depends on the availability of mathematical programming software, which is incorrect on two grounds: (1) portfolio optimizations can be implemented without mathematical programming, and (2) one does not need "hundreds or even thousands of assets" to optimize a portfolio.   For example, Statman (1987) indicates that the benefits of diversification may not be achieved if fewer than 30 stocks are used, and Newbould and Poon (1993) raise the estimate considerably to 80.   It is true that some portfolios have more securities than recommended by the previous studies, and it seems the reason has little to do with economics and more to do with the pressure investment managers feel to add certain stocks and industries. Lakonishok, Schleifer, and Vishny (1992) show that performance of the money management industry leaves much to be desired, and Wagner and Banks (1992) have suggested that trading costs, which are also related to having many securities to adjust to optimal positions, may be to blame.

We have covered our objective of reviewing portfolio selection models.   The next subsection will briefly outline the specifications that are suitable for reformulation in approximate form and seem to have practical value.

## C. The Selected Models

It is time to present the models we will formulate in approximate form. They are the following:

- *Model 1:* Mean-variance, excluding short sales and including investment in the risk-free rate. This is the model presented in the top half of Exhibit 5.7.
- *Model 2*: Single-index, excluding short sales and including investment in the risk-free rate. This is the model presented in the bottom half of Exhibit 5.7.   In its

exact, nonapproximate form, this model is equivalent to the single-index EGP algorithm.

**Exhibit 5.7**
**Selected Models**

The problem is:

Minimize $\sigma p^2$

Subject to a) $(\Sigma\ wi\ ri) + (1 - \Sigma\ wi)\ rf = rp$

Lagrangian $= [-\sigma p^2] - \lambda 1[\Sigma\ wi\ ri + (1 - \Sigma\ wi)\ rf - rp]$

## Model 1: Mean-variance, No short sales, rf included. (**)

| $\sigma 11$ | $\sigma 12$ | $\sigma 13$ | ... | $\sigma 1k\text{-}s$ | $r1\text{-}rf$ | | w1 | | 0 (*) |
|---|---|---|---|---|---|---|---|---|---|
| $\sigma 21$ | $\sigma 22$ | $\sigma 23$ | ... | $\sigma 2k\text{-}s$ | $r2\text{-}rf$ | | w2 | | 0 |
| ... | ... | ... | ... | ... | ... | | ... | = | ... |
| $\sigma k\text{-}s,1$ | $\sigma k\text{-}s,2$ | $\sigma k\text{-}s,3$ | ... | $\sigma k\text{-}s,k\text{-}s$ | $rk\text{-}s\text{-}rf$ | | wk-s | | 0 |
| $r1\text{-}rf$ | $r2\text{-}rf$ | $r3\text{-}rf$ | ... | $rk\text{-}s\text{-}rf$ | 0 | | $\lambda 1$ | | Rp-rf |

A                                        X = C

## Model 2: Single-index, No short sales, rf included. (*) (**)

| $(\beta 1^2\sigma m^2 + \sigma e1^2)$ | $\beta 1\beta 2\sigma m^2$ | $\beta 1\beta 3\sigma m^2$ | . | $\beta 1\beta k\text{-}s\sigma m^2$ | $r1\text{-}rf$ | | w1 | | 0 |
|---|---|---|---|---|---|---|---|---|---|
| $\beta 2\beta 1\sigma m^2$ | $(\beta 2^2\sigma m^2 + \sigma e2^2)$ | $\beta 2\beta 3\sigma m^2$ | . | $\beta 2\beta k\text{-}s\sigma m^2$ | $r2\text{-}rf$ | | w2 | | 0 |
| ... | ... | ... | . | ... | ... | | ... | = | 0 |
| $\beta k\text{-}s\beta 1\sigma m^2$ | $\beta k\text{-}s\beta 2\sigma m^2$ | $\beta k\text{-}s\beta 3\sigma m^2$ | . | $(\beta k\text{-}s^2\sigma m^2+\sigma ek\text{-}s^2)$ | $rk\text{-}s\text{-}rf$ | | wk-s | | 0 |
| $r1\text{-}rf$ | $r2\text{-}rf$ | $r3\text{-}rf$ | . | $rk\text{-}s\text{-}rf$ | 0 | | $\lambda 1$ | | Rp-rf |

A                                        X = C

Notes:

(*) After some simplification.

(**) Under no short sales, the tangent or unlevered portfolio must not contain securities with negative weights. Therefore, those "s" securities must be excluded from the original basis of "k" securities. This can be done through successive iterations of the basis reduction method, finding in each iteration the values for (rp-rf) that make the sum of optimal weights equal to one.

---

The first model selected is an extension of the model presented in Exhibit 5.2, which depicted the mean-variance model without short sales and specifying the required return on the portfolio. The second model selected is an extension of the model depicted in Exhibit 5.5, which summarized the single-index model without short sales and specifying the required portfolio return.

Models 1 and 2 calculate portfolios along the linear efficient frontier EF2 depicted in Exhibit 5.3. In these two models, the sum of optimal weights can be larger, equal, or smaller than one, depending on whether our required return is such that we borrow, allocate all our funds in the tangent or unlevered portfolio, or we lend funds, respectively. We can calculate the tangent portfolio by iterating the basis reduction method, as it is noted in Exhibit 5.7. (The GOAL SEEK facility in Excel's Solver can easily find in each iteration the value of (rp-rf) that makes the sum of optimal weights equal to one. We are avoiding weight rebalancing techniques because those techniques could force us to use extended interval arithmetic in the approximate specifications. This is not difficult to do, but may complicate our optimization "pictures" and there is no need for us to do so.)

At this point the discerning investor may ask: (1) Why not require that our sample yield market returns, at market risk, for the tangent portfolio?, and (b) If we have trusted in portfolio theory up to this point, why not extend our trust and invest in an index fund and a money market fund? We think these are questions the investor may have to face directly, in part because we enter into waters uncharted by the theory, and in part because the investor is using his/her own money. All we can do as analysts is select those parts of the theory that may be better fitted to provide support to investors. There is no right answer to question (2). With respect to question (1), a good sample should bring the results close to actual market risk and variance, if the index we use is a good approximation of the market itself. In this context, we recommend excluding short sales—at least the standard and Lintner versions—and including the risk-free rate. I am currently examining the implications of formulating the realistic short sales model in interval form. Inclusion of market returns and variance, even if indirectly through regression technique, can only benefit the exercise by further filtering the data concerning securities. The risk-free rate should be included: it is a fact of the investment reality and plays a crucial role in portfolio financing (lending and borrowing) and trading decisions as an alternative to hedging via short selling. Moreover, cash is the fluid that lubricates financial management.

A minor point to note is that we are not using Sharpe's diagonal model. There is no computational advantage to having zeroes in matrices when they are to be expressed in interval form. Matrices with zeroes may be more prone to ill conditioning, since the zeroes might represent missing links among equations.

A more important issue concerns the variance of optimal portfolios. Let MV-$\sigma p2$ and SI-$\sigma p2$ represent the portfolio variance obtained by the mean-variance and single-index models, respectively. They are related through the following equation:

(5)            $$MV\text{-}\sigma p^2 = SI\text{-}\sigma p^2 + \sum \sum_{i\ j,\ i \neq j} wi\ wj\ cov\ (\varepsilon i,\ \varepsilon j)$$

As Tucker et al. (1994, p. 114) note, "Hammer and Phillips (1992) estimated the cross-sectional residual covariances for 1,653 firms using monthly data from

1980 to 1989. They find that the single-index assumption of independent residuals is violated, with 72% of the 1.4 million cross-sectional residual covariance statistics positive." This selection of models includes both SI and MV to allow readers to judge the severity of this issue for themselves. Note that rather than estimating first the remainder term, it is best to calculate both portfolio variances—for portfolios with identical samples and portfolio returns—and observe whether there are any major differences between them.

Note also that short sales in our mean-variance optimizations are not included, nor are portfolios that yield returns below the risk-free rate, both of which should eliminate the "error maximization" tendency of mean-variance optimizations that do otherwise.

## 3. PRACTICAL PORTFOLIO SELECTION MODELS

While there is a generalized trend toward quantitative methods in the investment industry, it is fair to say that portfolio theory is not yet used with the frequency we would expect. Carter and Auken note the following:

(S)urprisingly, few IM [investment managers] apply portfolio analysis (30%), nor is there wide use of options and futures strategies (24% and 20%, respectively). For investment firms with a large number of securities, portfolio analysis may appear to be somewhat redundant, as diversification effects can cause return patterns to be similar to those of the market. . . . While a large proportion of investment managers appear to be using fundamental analysis, the use of portfolio and technical analysis and options and futures strategies is far less common. (1990, pp. 83, 84)

This section will introduce practical portfolio selection models. That is, models where the optimization is performed under reasonable assumptions that incorporate as many realistic features as possible—for example, presence of a risk-free and market rate—and are expressed in approximate form.

### A. Motivation and Data

Why errors? We should rather say: Why not errors? It should be obvious why we should always include errors (approximate equations systems) in portfolio optimization after our review of portfolio theory. First, portfolio optimization models are representations of a very complex piece of the investment reality; second, these representations are made simple by choosing to treat the problem statically, rather than dynamically; third, implicit and explicit assumptions simplify the problem perhaps at the cost of lost relevance; fourth, we have excluded many alternative securities, real investments, and markets from the final portfolio optimizations; fifth, the meager set of statistical indicators we use is subject to considerable sampling, estimation, and measurement error; sixth, ultimately, the objects we use—Jacobian matrices—have only so much "room for error." Note also how much of the uncertainty we face in portfolio management extends beyond that of a purely probabilistic (frequency-based) nature.

Going over each specific reason for preferring the approximate to the customary "exact" form would replicate much of the material presented already. However, it seems useful to highlight a few specific reasons, which we will do in schematic form.  First, there is the issue of model selection—choosing between mean-variance and the single-index specification, which cannot be settled even if we agreed that these models were reasonable representations of the investment reality.  When we select the mean-variance model, we further assume that two parameters (mean and variance) describe the statistical behavior of the variables involved, and that structural change (changing means and variances) is not a problem.  When we select the single-index model, additional issues are brought about by inclusion of the concepts of a market portfolio and market return. (It should be clear that common regressions also assume the sufficiency of two-parameter approximations.)  Harrington (1987) presents a thorough analysis of the many practical problems we will encounter when trying to implement the single-index model, many of which relate to assessing the likely behavior of the market, which is a very difficult issue to tackle—see Merton (1980).

Independently of the model selected, the analyst must make important sampling decisions (length of the interval, frequency of the data, and so on). Some of these decisions are seemingly trivial, such as the decision to use biased—but consistent—estimates for variance estimates ($\sigma2 = \Sigma$ (xi – mean)2/n-1), or maximum likelihood estimates ($\sigma2 = \Sigma$ (xi – mean)2/n).  The argument is clear; using "n" provides the maximum likelihood unbiased estimate when n is large, but does not converge to the true estimate when n is small.  On the other hand, (n–1) provides a biased estimate that, nonetheless, converges to the true value as n becomes large (consistency).  As a general rule, with small samples, one is better off using (n–1) because the chance of underestimating true risk (and overestimating expected returns) is diminished.  But in making a decision about this seemingly minor issue, two important additional considerations emerge: (1) sometimes we may not have large samples, either because the company is new or because it has gone through special processes of unlikely repetition; (2) optimal weights are very sensitive to even very small changes in the data; and (3) many digits in the optimal weight estimates become important when we allocate, say, 10 million dollars in a company.

The frequency-based statistical nature of the tools we use is a cause for concern in terms of accuracy.  For example, the normal existence of firms (and people) can be understood as a flowing of special circumstances and unending change.  However, the very nature of classical (or frequency-based) probability presumes absence of change and "special" situations for any given period of reference.  There is little doubt that large market shocks invalidate much of the predictive power of descriptive statistics, see Levy and Yoder (1989), but even on a day-to-day basis one must wonder whether the "business as usual" of money managers is dealing with irregularities rather than managing the type of regularities and stable phenomena presupposed by classical probability.

**Exhibit 5.8**
**Companies Used in This Monograph**

| Name | Ticker | Market | Activity |
|---|---|---|---|
| Apple Computer | AAPL | NASDAQ | Computer manufacturing |
| ABM Industries | ABM | NYSE | Real estate investment |
| Autodesk | ADSK | NASDAQ | Computer aided design |
| Acuson | ACN | NYSE | Med. ultrasound Equip. |
| Adobe Systems | ADBE | NASDAQ | Computer software |
| Adaptec | ADPT | NASDAQ | Networking products |
| Altera | ALTR | NASDAQ | Integrated circuits |
| Applied Materials | AMAT | NASDAQ | Semiconductors |
| Adv. Micro Devices | AMD | NYSE | Semiconductors |
| Amdahl | AMH | AMEX | Computer hardware |
| Alza | AZA | NYSE | Drug delivery |
| BankAmerica | BAC | NYSE | Commercial banking |
| Franklin Resources | BEN | NYSE | Financial services |
| Bio-Rad Labs A | BIOA | AMEX | Diagnostic products |
| Cadence Design Sys. | CDN | NYSE | Electronic design |
| Chiron | CHIR | NASDAQ | Biopharmaceuticals |
| Chevron | CHV | NYSE | Oil products |
| Clorox | CLX | NYSE | Consumer goods |
| CMC Industries | CMIC | NASDAQ | Electronic manufacturing |
| CNF Transportation | CNF | NYSE | Freight transportation |
| Market index | SPX | | Standard & Poor's 500 |

- NYSE is the New York Stock Exchange, NASDAQ is the National Association of Security Dealers, and AMEX is the American Stock Exchange.
- Stock returns calculated as natural logarithm from price relatives, $rt=\ln(pt/pt-1)$, excluding dividends, from closing prices for 60 monthly observations ( 5 years) covering the period September 1992–August 1997.
- A risk-free rate estimate of 5% is used throughout.

We have already made the case for using approximate equations using the old, plain measurement error argument. Once again, note that the actual investment problem we are solving in portfolio management is forward-looking, while all the estimates we build refer to the past. In the stock market, historical estimates will always be, at best, only approximations of future values. The estimates will likely be biased as long as the future is different from the past. A person assuming "rational expectations" may have the romantic tendency to believe that people can manufacture unbiased predictors of future values, but we know of no realistic investing model that can do that.

Lastly, note that every research initiative in portfolio management stems from recognizing the many and daunting nature of errors in our methods. We use these methods because, very simply stated, we have no better alternatives. Note also that some of the aforementioned important research questions perhaps cannot be addressed unless an approximate framework is used. In particular, issues concerning nonstationarity, estimation ranges for means and variances, expectations ranges for market returns and desired returns on the portfolio, and sampling issues concerning the estimates need to include considerations for allowable errors that portfolio Jacobians—matrices from first order conditions— can absorb without becoming critically ill conditioned. For example, there is little use in calculating Chi-square ranges for variances if we do not first study what types of ranges would cause critical ill conditioning in the optimization. Our experience is that ranges for variance estimates are several times larger than those allowable by the optimization matrices. In addition to Michaud (1998), additional references concerning nonstationarities and portfolio choice are Barry and Winkler (1976), Barry (1974) and Buser (1977).

It is now time to apply approximate equations to the portfolio problem. We will do so by adding errors to the matrices obtainable from the first order conditions (FOC) of each optimization, that is, the Jacobians. Of course, it would be better to develop a portfolio management model based on interval-valued real numbers from scratch. Our strategy represents a compromise between what is done now, to work with presumably "exact" specifications, and what could be done in the future: use full-fledged interval-based portfolio models. Our strategy is similar to what is routinely done in Econometrics, where we add white noise to the exact specifications.

Exhibits 5.8, 5.9, and 5.10 present the companies we used in our application of approximate systems to practical portfolio optimization and the summary statistics for mean-variance and the single-index models, respectively. A rate of 5% has been assumed for the risk-free rate.

## B. Results

This is the most important segment of this chapter. Improving portfolio optimization depends on a two-stage sequence. We must first obtain the "exact" optimal (unlevered or tangent) portfolios described in models 1 and 2 of Exhibit 5.7. These portfolios are the most interesting from a practical point of view because they are obtained under a no short-sales condition and include the risk-free rate. Exhibits 5.11 and 5.12 show that both the basis reduction method and the technique to calculate unlevered portfolios work. Note also that the sum of optimal weights equals one, even without including an explicit restriction. The tangent portfolio implies no borrowing (a sum of weights larger than one) or lending (a sum of weights less than one), and can be calculated using Excel's "goal seek" feature. Not having an explicit restriction is also helpful in calculating approximate weights because it lowers the dimension of the problem.

**Exhibit 5.9**
**Mean-Variance Estimates**

| Ticker | AAPL | ABM | ADSK | ACN | ADBE | ADPT | ALTR | AMAT | AMD | AMH | AZA | BAC | BEN | BIOA | CDN | CHIR | CHV | CLX | CMIC | CNF |
|---|---|---|---|---|---|---|---|---|---|---|---|---|---|---|---|---|---|---|---|---|
| AAPL | 0.01735 | | | | | | | | | | | | | | | | | | | |
| ABM | 0.00164 | 0.00354 | | | | | | | | | | | | | | | | | | |
| ADSK | 0.00313 | -0.00043 | 0.01433 | | | | | | | | | | | | | | | | | |
| ACN | 0.00139 | 0.00151 | 0.00161 | 0.01108 | | | | | | | | | | | | | | | | |
| ADBE | 0.00847 | -0.00004 | 0.00603 | -0.00164 | 0.02537 | | | | | | | | | | | | | | | |
| ADPT | 0.00454 | 0.00054 | 0.00391 | 0.00305 | 0.00269 | 0.01417 | | | | | | | | | | | | | | |
| ALTR | 0.00151 | -0.00154 | 0.00179 | 0.00288 | -0.00110 | 0.00506 | 0.02280 | | | | | | | | | | | | | |
| AMAT | 0.00117 | 0.00017 | 0.00073 | 0.00213 | 0.00610 | 0.00302 | 0.00942 | 0.01569 | | | | | | | | | | | | |
| AMD | 0.00242 | -0.00030 | 0.00516 | 0.00290 | 0.00190 | 0.00199 | 0.00663 | 0.00654 | 0.02241 | | | | | | | | | | | |
| AMH | 0.00285 | 0.00143 | 0.00443 | 0.00199 | 0.00559 | 0.00170 | 0.00237 | 0.00302 | -0.00012 | 0.01922 | | | | | | | | | | |
| AZA | 0.00155 | 0.00046 | 0.00108 | 0.00161 | -0.00299 | 0.00262 | 0.00059 | 0.00020 | -0.00071 | 0.00017 | 0.01161 | | | | | | | | | |
| BAC | -0.00075 | 0.00035 | 0.00127 | 0.00078 | -0.00020 | 0.00130 | 0.00292 | 0.00293 | -0.00007 | 0.00189 | 0.00249 | 0.00493 | | | | | | | | |
| BEN | -0.00064 | -0.00014 | 0.00305 | 0.00217 | -0.00009 | 0.00212 | 0.00611 | 0.00571 | 0.00092 | 0.00161 | 0.00162 | 0.00302 | 0.00637 | | | | | | | |
| BIOA | 0.00036 | 0.00042 | 0.00375 | 0.00162 | 0.00278 | 0.00166 | 0.00119 | 0.00407 | 0.00106 | 0.00378 | 0.00178 | 0.00100 | 0.00103 | 0.00949 | | | | | | |
| CDN | 0.00312 | -0.00010 | 0.00445 | 0.00185 | 0.00481 | 0.00484 | 0.00446 | 0.00222 | -0.00106 | 0.00447 | 0.00108 | 0.00096 | 0.00081 | 0.00374 | 0.01763 | | | | | |
| CHIR | 0.00072 | 0.00149 | 0.00016 | 0.00350 | -0.00134 | -0.00043 | 0.00213 | 0.00066 | -0.00288 | 0.00011 | 0.00144 | 0.00001 | 0.00214 | 0.00046 | 0.00076 | 0.01092 | | | | |
| CHV | 0.00024 | 0.00014 | 0.00044 | -0.00027 | 0.00108 | 0.00108 | 0.00063 | 0.00063 | -0.00025 | 0.00122 | 0.00009 | 0.00055 | 0.00100 | 0.00021 | -0.00084 | 0.00002 | 0.00181 | | | |
| CLX | -0.00119 | 0.00026 | 0.00029 | 0.00096 | -0.00288 | -0.00080 | 0.00235 | 0.00089 | 0.00001 | -0.00114 | 0.00156 | 0.00142 | 0.00168 | -0.00002 | -0.00134 | 0.00141 | 0.00002 | 0.00241 | | |
| CMIC | 0.00265 | -0.00024 | 0.00266 | 0.00117 | -0.00041 | 0.00201 | 0.00871 | 0.00533 | 0.00462 | 0.00318 | -0.00118 | 0.00026 | 0.00026 | 0.00140 | 0.00127 | 0.00226 | 0.00098 | 0.00058 | 0.01658 | |
| CNF | 0.00361 | 0.00093 | 0.00326 | 0.00082 | 0.00593 | 0.00277 | 0.00212 | 0.00259 | 0.00013 | 0.00119 | 0.00109 | 0.00069 | 0.00194 | 0.00033 | 0.00198 | 0.00051 | 0.00077 | -0.00024 | 0.00138 | 0.00770 |
| | | | | | | | | | | | | | | | | | | | | |
| Ri | -0.01600 | 0.01312 | 0.00632 | 0.00527 | 0.01397 | 0.03015 | 0.04077 | 0.04032 | 0.01490 | 0.00083 | -0.01012 | 0.01605 | 0.01949 | 0.01108 | 0.02650 | 0.00604 | 0.00831 | 0.01469 | 0.00176 | 0.01377 |
| Vai | 0.01735 | 0.00354 | 0.01433 | 0.01108 | 0.02537 | 0.01417 | 0.02280 | 0.01569 | 0.02241 | 0.01922 | 0.01161 | 0.00493 | 0.00637 | 0.00949 | 0.01763 | 0.01092 | 0.00181 | 0.00241 | 0.01658 | 0.00770 |
| Std | 0.13173 | 0.05952 | 0.11971 | 0.10525 | 0.15896 | 0.11905 | 0.15100 | 0.12527 | 0.14969 | 0.13862 | 0.10775 | 0.07020 | 0.07982 | 0.09739 | 0.13279 | 0.10448 | 0.04235 | 0.04913 | 0.12876 | 0.08772 |

**Exhibit 5.10**
**Single-Index Estimates**

| Ticker | ri | betai | vari | resvari |
|---|---|---|---|---|
| S&P500 | 0.013393 | 1 | 0.000871 | 0 |
| 1 AAPL | -0.011832 | 0.308219 | 0.017352 | 0.017270 |
| 2 ABM | 0.017287 | 0.346252 | 0.003542 | 0.003438 |
| 3 ADSK | 0.010489 | 1.341444 | 0.014330 | 0.012763 |
| 4 ACN | 0.009440 | 1.043432 | 0.011077 | 0.010128 |
| 5 ADBE | 0.018137 | 0.739909 | 0.025269 | 0.024792 |
| 6 ADPT | 0.034321 | 1.342239 | 0.014173 | 0.012603 |
| 7 ALTR | 0.053437 | 1.803797 | 0.022801 | 0.019966 |
| 8 AMAT | 0.044490 | 2.402583 | 0.015693 | 0.010664 |
| 9 AMD | 0.019065 | 0.280453 | 0.022408 | 0.022339 |
| 10 AMH | 0.004993 | 0.990257 | 0.019216 | 0.018361 |
| 11 AZA | -0.005951 | 1.457662 | 0.011609 | 0.009758 |
| 12 BAC | 0.020220 | 1.663024 | 0.004928 | 0.002518 |
| 13 BEN | 0.023657 | 1.850095 | 0.006372 | 0.003390 |
| 14 BIOA | 0.015248 | 1.021237 | 0.009486 | 0.008577 |
| 15 CDN | 0.030668 | 0.637411 | 0.017634 | 0.017280 |
| 16 CHIR | 0.010204 | 0.535127 | 0.010916 | 0.010666 |
| 17 CHV | 0.012482 | 0.555709 | 0.001811 | 0.001542 |
| 18 CLX | 0.018861 | 0.678634 | 0.002414 | 0.002013 |
| 19 CMIC | 0.005927 | 0.619373 | 0.016579 | 0.016245 |
| 20 CNF | 0.017939 | 0.819877 | 0.007696 | 0.007110 |

Let us start with the mean-variance specification. The initial sample contained 20 securities and its first optimization yielded the weights presented on the left-hand side of Exhibit 5.11. The optimal, unlevered, tangent portfolio contained nine securities, and appears on the right-hand side of the top of the same exhibit. The last portfolio of nine securities could be computed directly with quadratic programming. However, this is a book about simultaneous equations systems and they are used here for everything—even if it means developing new procedures and techniques in portfolio optimization, which also has been done. Exhibit 5.12 presents the optimization results for the single-index specification. The optimal, tangent, unlevered portfolio contains 11 securities, two more than the mean-variance model, although the last security in the single-index model has negligible weight. We could say that both models are comparable as to securities selected in the optimal portfolio. That the single-index and the mean-variance specifications share most securities should not be a surprise: the main diagonal is identical in both, and convergence requires the matrix to be

## Exhibit 5.11
## Optimization Results: Mean-Variance Model

| 1st Optimization | | 2nd Optimization | | 3rd Optimization | | QP-No SS-rf included | |
|---|---|---|---|---|---|---|---|
| Security | wi* | Security | wi* | Security | wi* | Security | wi* |
| 18 | 0.42740957 | 18 | 0.3385148906 | 18 | 0.3308802020 | 18 | 0.3308802020 |
| 17 | 0.33510919 | 17 | 0.2438593227 | 17 | 0.2314767578 | 17 | 0.2314767578 |
| 2 | 0.25656069 | 2 | 0.2058256693 | 2 | 0.1932361893 | 2 | 0.1932361893 |
| 15 | 0.14558061 | 15 | 0.1007896057 | 15 | 0.0935595072 | 15 | 0.0935595072 |
| 8 | 0.09371718 | 8 | 0.0560408577 | 7 | 0.0502245955 | 7 | 0.0502245955 |
| 7 | 0.06589836 | 7 | 0.0516207168 | 8 | 0.0467903888 | 8 | 0.0467903888 |
| 9 | 0.05421346 | 5 | 0.0365134573 | 5 | 0.0295410706 | 5 | 0.0295410706 |
| 20 | 0.05183676 | 9 | 0.0146178299 | 9 | 0.0181627599 | 9 | 0.0181627599 |
| 6 | 0.04471677 | 6 | 0.0081337748 | 6 | 0.0061285289 | 6 | 0.0061285289 |
| 5 | 0.03157906 | 20 | -0.0279031596 | | | | |
| 14 | 0.02782030 | 14 | -0.0280129652 | | | | |
| 16 | -0.00247389 | | | | | | |
| 12 | -0.01877668 | | | | | | |
| 10 | -0.04339933 | | | | | | |
| 13 | -0.05490880 | | | | | | |
| 4 | -0.05955670 | | | | | | |
| 3 | -0.06824410 | | | | | | |
| 19 | -0.08670094 | | | | | | |
| 1 | -0.09072264 | | | | | | |
| 11 | -0.10965888 | | | | | | |

| | | | | | | | |
|---|---|---|---|---|---|---|---|
| Sum wi | 1.000000 | | 1.000000 | | 1.000000 | | 1.000001 |
| k-s = | 20.000000 | | 11.000000 | | 9.000000 | | 9.000000 |
| Rp-rf | 0.027177 | | 0.017455 | | 0.01703103 | | 0.01703102 |
| yearly % | 32.61222278 | | 20.94611715 | | 20.43723627 | | 20.43722193 |
| sigma-p | 0.03891426 | | 0.03122910 | | 0.03061116 | | 0.03061113 |
| yearly sig | 46.69711749 | | 37.47492207 | | 36.73338762 | | 36.73335432 |
| Sharpe | 0.69837764 | | 0.55893691 | | 0.55636677 | | 0.55636689 |

## Exhibit 5.12
## Optimization Results: Single-Index Model

| 1st Optimization | | 2nd Optimization | | 3rd Optimization | | 4th Optimization | | QP-No SS-rf included | |
|---|---|---|---|---|---|---|---|---|---|
| Security | wi* | Security | wi* | Security | wi* | Security | wi* | Security | wi* |
| 18 | 0.341239 | 18 | 0.264560369 | 18 | 0.252335 | 18 | 0.251621289 | 18 | 0.251616226 |
| 2 | 0.238563 | 2 | 0.196335563 | 2 | 0.192489 | 2 | 0.192125257 | 2 | 0.192130367 |
| 17 | 0.16339 | 17 | 0.109895676 | 7 | 0.104242 | 7 | 0.104004117 | 7 | 0.104007246 |
| 8 | 0.136382 | 7 | 0.107579297 | 17 | 0.097111 | 17 | 0.096569957 | 17 | 0.09657391 |
| 7 | 0.134169 | 8 | 0.098102015 | 8 | 0.090063 | 8 | 0.089686825 | 8 | 0.089696691 |
| 6 | 0.11502 | 6 | 0.089776403 | 6 | 0.085904 | 6 | 0.085670919 | 6 | 0.085668481 |
| 15 | 0.098725 | 15 | 0.081653667 | 15 | 0.080229 | 15 | 0.080082907 | 15 | 0.080087808 |
| 20 | 0.069947 | 20 | 0.050334122 | 20 | 0.046219 | 20 | 0.046026793 | 20 | 0.046014637 |
| 9 | 0.045709 | 9 | 0.03818608 | 9 | 0.037684 | 9 | 0.037620899 | 9 | 0.037621889 |
| 13 | 0.045489 | 5 | 0.017484388 | 5 | 0.016412 | 5 | 0.016356475 | 5 | 0.016357487 |
| 5 | 0.023295 | 16 | 0.002050195 | 16 | 0.000293 | 16 | 0.000234563 | 16 | 0.000225268 |
| 14 | 0.013067 | 14 | 0.001182647 | 14 | -0.002982 | | | | |
| 16 | 0.007314 | 13 | -0.006953181 | | | | | | |
| 12 | 0.006396 | 12 | -0.050187243 | | | | | | |
| 19 | -0.021321 | | | | | | | | |
| 4 | -0.038584 | | | | | | | | |
| 10 | -0.039206 | | | | | | | | |
| 3 | -0.042357 | | | | | | | | |
| 1 | -0.091607 | | | | | | | | |
| 11 | -0.205631 | | | | | | | | |

| | | | | | | | | | |
|---|---|---|---|---|---|---|---|---|---|
| Sum wi | 1 | | 1 | | 1 | | 1 | | 1.00000001 |
| k-s = | 20 | | 14 | | 12 | | 11 | | 11 |
| Rp-rf | 0.033851 | | 0.022095371 | | 0.021921 | | 0.021891912 | | 0.021892351 |
| yearly % | 0.406206 | | 0.265144447 | | 0.263054 | | 0.262702939 | | 0.262708211 |
| sigma-p | 0.053177 | | 0.039808693 | | 0.039618 | | 0.039566639 | | 0.039567433 |
| yearly sig | 0.638119 | | 0.477704318 | | 0.475422 | | 0.474799665 | | 0.474809193 |
| Sharpe | 0.636569 | | 0.555038833 | | 0.553306 | | 0.553292174 | | 0.553292175 |

**Exhibit 5.13**
**Approximate Equations: Mean-Variance Model**

Maximum error = 0.000242562

| Optimal    Weights | Error 1 = | 0.00023 | Textbook | Error 2 = | 0.000194 |
|---|---|---|---|---|---|
| Mean-Variance | IC | 0.95 | Weights | IC | 0.8 |
| Security    Ticker | Max E1 | Min E1 | Cent-Form | Max E2 | Min E2 |
| 2 ABM | 0.503806 | -0.422336 | 0.19323619 | 0.419567 | -0.178457 |
| 5 ADBE | 0.08466 | -0.095976 | 0.02954107 | 0.131111 | -0.0694 |
| 6 ADPT | 0.117946 | -0.255856 | 0.00612853 | 0.093255 | -0.146498 |
| 7 ALTR | 0.236979 | -0.387336 | 0.0502246 | 0.180164 | -0.204688 |
| 8 AMAT | 0.240732 | -0.163598 | 0.04679039 | 0.174204 | -0.112589 |
| 9 AMD | 0.210501 | -0.081438 | 0.01816276 | 0.142347 | -0.047241 |
| 15 CDN | 0.398474 | -0.064338 | 0.09355951 | 0.280501 | -0.011698 |
| 17 CHV | 1.147433 | -0.242842 | 0.23147676 | 0.793044 | -0.187203 |
| 18 CLX | 1.435805 | -0.176275 | 0.3308802 | 0.986384 | -0.09783 |

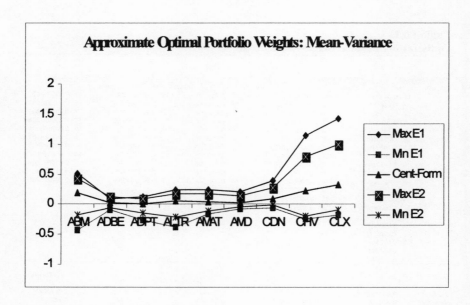

**Exhibit 5.14**
**Approximate Equations: Single-Index Model**

Maximum error = 0.000318462

| Optimal Single-Index | Weights | Error 1 = IC | 0.00030254 0.95 | Textbook Weights | Error 2 = IC | 0.0002548 0.8 |
|---|---|---|---|---|---|---|
| Security | Ticker | Max E1 | Min E1 | Cent-Form | Max E2 | Min E2 |
| 2 | ABM | 0.514645 | -0.1376851 | 0.19212526 | 0.428029 | -0.069669 |
| 5 | ADBE | 0.062577 | -0.0319265 | 0.01635647 | 0.049704 | -0.025559 |
| 6 | ADPT | 0.202308 | -0.0571382 | 0.08567092 | 0.174646 | -0.0274791 |
| 7 | ALTR | 0.200229 | -0.0143326 | 0.10400412 | 0.182206 | 0.0096774 |
| 8 | AMAT | 0.238727 | -0.1108765 | 0.08968682 | 0.208255 | -0.0623036 |
| 9 | AMD | 0.087046 | -0.023388 | 0.0376209 | 0.078646 | -0.0092165 |
| 15 | CDN | 0.160562 | -0.0119776 | 0.08008291 | 0.145603 | 0.0042065 |
| 16 | CHIR | 0.10317 | -0.0974714 | 0.00023456 | 0.075838 | -0.0780535 |
| 17 | CHV | 0.663371 | -0.7026443 | 0.09656996 | 0.581442 | -0.4977848 |
| 18 | CLX | 0.733595 | -0.3081306 | 0.25162129 | 0.635878 | -0.1325577 |
| 20 | CNF | 0.180712 | -0.1358263 | 0.04602679 | 0.159795 | -0.0990495 |

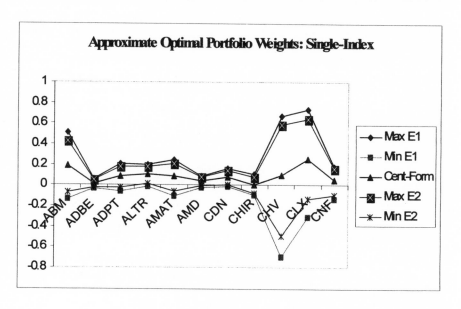

diagonally dominant, which means that main effects (main diagonal) are more important than cross effects (off diagonal).

The right-hand-side of Exhibits 5.11 and 5.12 shows the differences between the iterative and quadratic programming procedures. For the mean-variance model the final (no short sales) optimizations are identical; however, for the single-index there is a (very small) loss of precision, perhaps due to a difference in precision between iterative and quadratic programming solutions. The SES approach yields identical results to quadratic programming for the initial (short sales allowed) optimization. The differences are very small, and their summation appears to be random and negligible.

Exhibits 5.13 and 5.14 present the second stage of our strategy for improving portfolio optimization in practice: the application of approximate procedures to our portfolio optimizations. The first line in each of these exhibits shows the maximum allowable error that would provoke critical ill-conditioning if added/subtracted from each element of the matrix system. The middle portion shows uncertainty ranges for optimal portfolio weights when errors are applied to the Jacobian: the larger error is associated with an ill-conditioning value of 0.95, and the smaller error with an ill-conditioning factor of 0.80. The lower portion of each exhibit shows a graph depicting the conventional "exact" optimal solution, which includes the optimal weights we see in textbooks (represented by triangle symbols in these two exhibits).

Approximate weights are the initial, center value, plus its uncertainty range. This is the first time these types of ranges have been obtained for portfolio optimizations. They challenge us to re-evaluate some fundamental notions about portfolio management such as the accuracy of portfolio construction, the distinguishability of capital market theories, the relevance and reliability of optimal positions, and even some notions on the regulation of security holdings. The next section will highlight some issues concerning portfolio approximations.

Note that what I have done is to use approximations to migrate from the exact, ideal, or abstract models, to the real, concrete, or simply practical versions of our representations of an optimal portfolio. This is the lesson from this section: given our limited information, optimal portfolio weights can only be expected to lie within a range.

## C. Analysis

We could make many comments based upon the results presented in Exhibits 5.13 and 5.14. Perhaps the best way to start is by noting what should have been obvious to the reader: uncertainty ranges in portfolio optimization are very large. In other words, information in the problem is limited and wide ranges of possible values exist for the resulting "exact" optimal weights. Portfolio optimizations are extremely imprecise.

The imprecision is so large it may convince some readers to never use portfolio optimization in practice until portfolio analysis changes significantly:

**Exhibit 5.15**
**Confidence Intervals for Means and Variances**

| n | 60 | | | | | | | | |
|---|---|---|---|---|---|---|---|---|---|
| Ticker | ABM | ADBE | ADPT | ALTR | AMAT | AMD | CDN | CHV | CLX |
| Number | 2 | 5 | 6 | 7 | 8 | 9 | 15 | 17 | 18 |
| wi* | 0.193176 | 0.029528 | 0.006178 | 0.050175 | 0.046831 | 0.018178 | 0.093556 | 0.231509 | 0.33087 |
| | | | | | | | | | |
| Returns | 0.013121 | 0.01397 | 0.030154 | 0.04927 | 0.040323 | 0.014898 | 0.026501 | 0.008315 | 0.014695 |
| Std error | 0.001694 | 0.001804 | 0.003893 | 0.006361 | 0.005206 | 0.001923 | 0.003421 | 0.001073 | 0.001897 |
| Upper lim | 0.016441 | 0.017505 | 0.037784 | 0.061737 | 0.050526 | 0.018668 | 0.033206 | 0.010419 | 0.018413 |
| Lower lim | 0.009801 | 0.010435 | 0.022524 | 0.036803 | 0.03012 | 0.011128 | 0.019795 | 0.006211 | 0.010976 |
| | | | | | | | | | |
| Variances | 0.003542 | 0.025269 | 0.014173 | 0.022801 | 0.015693 | 0.022408 | 0.017634 | 0.001811 | 0.002414 |
| Std error | 0.000647 | 0.004613 | 0.002588 | 0.004163 | 0.002865 | 0.004091 | 0.003219 | 0.000331 | 0.000441 |
| Upper lim | 0.00481 | 0.034311 | 0.019245 | 0.03096 | 0.021309 | 0.030427 | 0.023944 | 0.002459 | 0.003278 |
| Lower lim | 0.002275 | 0.016226 | 0.009101 | 0.014642 | 0.010077 | 0.014389 | 0.011324 | 0.001163 | 0.00155 |

the evidence supports that assessment. The stunning lack of accuracy in portfolio optimization stems from the weakness of statistical indicators, which are so imprecise as to make them nearly irrelevant. Therefore, one could say there is a strong case to be made against using portfolio theory in practice as long as the optimization model is based exclusively on statistical estimates. Anyone with moderate exposure to statistical analysis and sampling would have been able to warn us about this before running the optimizations. For example, Exhibit 5.15 shows the standard errors associated with average returns and variances, and the 95% confidence interval obtainable by adding 1.96 standard errors to those estimates. These confidence intervals were obtained under the assumption of normality. A Chi-square test could have been used for variances, but as the degrees of freedom increase above 30 (59 in our case), the Chi-square approaches normal. If we drop the normality assumption, the intervals would provide only about 75% confidence via Chevisev's theorem. The ill-conditioning errors we used were the largest allowable by the Jacobians and are still much smaller (about 10 times smaller) than those provided by statistical procedures. This is important: the matrices we use in portfolio optimization are unable to accept the ranges implicit in the statistical procedures we perform prior to the optimization.

The major implication is that as long as our optimizations are based on statistical indicators, we may lack the informational power to settle the very critical theoretical hypotheses whose determination we take for granted. Some examples of these hypotheses are: (1) capacity of mean-variance analysis to help investors increase their wealth, (2) role of the market index in creating better portfolios, (3) ability to construct portfolios that establish a clear distinction between short sales and no-short sales, and (4) ability to incorporate the risk-free rate effectively in portfolio management.

But, as usual, just as the risk of collapse lays in the midst of construction, the promise of developing better portfolio analysis lays in the midst of destruction—what has been done in this section of the study.

First, uncertainty ranges could have been narrowed by doing some or all of the following:

1.  Including a $\Sigma wi = 1$ constraint in each mixed-integer optimization. Maximum and minimum values for wi* were independently calculated of whether the sum of weights would equal one. Doing so made sure that the most extreme points in the underlying polyhedron were identified, and revealed the mathematical properties of the objects without constraining them—artificially—to a priori economic restrictions.

2.  Lowering allowable error in the calculation of ranges of uncertainty. It is true these errors are likely to be at the very high end of the spectrum of realistic error estimates; but this revealed "worst case scenarios," in order to "prepare for the worst" even though our exercise has been "hoping for the best." Practitioners and scholars will find out about practical and helpful values for those maximum errors in the near future.

3.  By carefully filtering the sample to present a "well-behaved" case. I did not want to do any of that, which is not uncommon when authors present technical and theoretical developments. I simply went to a sample of companies in which I was interested, the San Francisco Bay Area that includes companies from the famous "Silicon Valley," and selected the first 20 based on their alphabetical order.

Therefore, it should be clear that in no way have the techniques and procedures we presented been "helped." Each of the three strategies mentioned above are likely to highlight the usefulness of approximate methods in portfolio management. Hopefully substantial empirical research on the methods introduced in this chapter and monograph will confirm these expectations of such practical usefulness.

Second, including ranges of uncertainty is the first step to linking portfolio optimization to expectations formation which, in turn, is essential to develop models that are well suited to individual cognitive and decision-making characteristics. This is a major step forward. Third, we have learned a great deal about portfolio optimization itself and a way to model the dual nature of the variables involved, given that in this problem each security can be bought or sold short, with different expected values in each case, see Tarrazo and Alves (1998b). These approximate procedures are one way—a feasible, yet sophisticated and sound way—to address the combinatorial complexity present in the investing decision (see Fogler, 1993).

There has been progress in yet another area. Practical means, above all, feasible. Every procedure carried out can be performed with modest informational and computational resources. The data employed can be obtained from the Internet at no cost, and processed with spreadsheet software. Portfolio theory was born to help individual investors and I believe that progress in portfolio analysis should respect the overriding goal of helping the individual investor.

What is next? The first suggestion for further study presented in this chapter centers on substantial empirical research and testing of approximate methods. This empirical research could be accompanied by studies on the structure of the problem using perhaps fundamental (accounting and financial) methods and, thus, move beyond purely statistical indicators. Qualitative methods could also complement the development of approximate quantitative methods.

## SUMMARY

Stephen Ross, one of the most influential authors of modern times, noted at the end of the 1970s: "The area of approximation contains an important class of problems with genuine empirical significance" (1977, p. 183). This chapter has applied approximate methods to portfolio analysis.

It started by surveying the general foundations of the investment decision (investors, markets, and securities), and by thoroughly and critically analyzing portfolio optimization methods from both mathematical and economic angles. The in-depth investigation revealed two models of theoretical significance and practical potential—what would be called the best portfolio selection has to offer. Then a sample of companies was selected and approximate methods were applied in a rigorous and strict manner to "test-drive" these approximate methods under hard, if not extreme, conditions. Analysis of results indicated the low informational power of statistical indicators that (1) are a staple of portfolio optimization, (2) may not help us assess important theoretical foundations of portfolio optimization (for example, role of the risk-free and market rates, short sales, and so on), and (3) might make current portfolio procedures unsuitable for practical purposes. However, approximate methods fit human cognition and the way we think. It is the proper way to manage the mathematical objects in portfolio analysis, and corresponds to the quality of available indicators, given the current state of knowledge concerning financial markets and security analysis. This chapter also suggested strategies to narrow uncertainty ranges for approximate optimal weights.

Approximate financial planning seems more likely to improve financial practice than approximate portfolio methods because in the former one can make uncertainty ranges narrower by simply studying historical financial statements. Approximate methods fare much better in both financial planning and portfolio selection than in macroeconomic analysis, where we seem to have been bent on developing constructs of little practical use.

In sum, approximate methods have potential to significantly change not only portfolio optimization but also the entire portfolio management process. Therefore, it is easy to conclude that in a not-so-distant future, securities will be selected with approximate methods, optimal portfolios expressed in interval form, and portfolio managers judged in terms of risk-return intervals. This is, after all, what looks extremely reasonable to us today, and what appears as the brightest future of portfolio analysis.

## APPENDIX: A NOTE ON THE LITERATURE OF PORTFOLIO ANALYSIS

Portfolio management is a combination of complex elements, some of which require undivided attention from the analyst, researcher, or practitioner. Some of these elements are: statistical analysis, optimization, consumer theory, macroeconomics, and finance. In addition, portfolio analysis can be approached from each of these angles to clarify and enhance the role played by each specific element. On top of this we must assimilate material regarding the investments industry and its regulations, even security- and exchange-specific detail in some cases. Investors also handle industries, companies, and domestic and foreign markets, which gives an idea of the cognitive and organizational skills that professional investment management demands. In this case, I have stressed the numerical element of portfolio management, which I believe also helps to integrate other elements for practical decision-making. The analysis has tried to streamline the presentation so it would not be burdened with customary references to original sources, follow-up references to textbooks, where the same material can be easier to access and benefits from modern developments, references to practitioners' experience with theoretical material, or references to explanations of technical details.

This appendix attempts to strengthen the bibliographical support to the analysis in a way that is helpful to the intended audience of practitioners, researchers, and self-students of portfolio analysis. I also believe that portfolio management is the door into quantitative investment management, which is the future, if not yet the present, of money management. Therefore, the rest of this appendix will honor the duty to acknowledge some of the material presented and highlight those references that will be useful to our audience.

Markowitz's (1952) study is considered the seminal work in portfolio analysis, which evolved into a well-worth-reading book (1959). Tobin (1958) is one of the best studies we have ever read. Monetary economics is an extremely active area of research and James Tobin, a Nobel prize winner in economics, jumped into the study of portfolio theory because it offered promise for his area of concern. In his 1958 article, Tobin provided the following: a theory for the demand of money, a theory for the demand of treasuries (as they can also represent cash), the one-fund separation theorem, parts of the theory of portfolio optimization in the presence of a risk-free asset, and a theory of optimal portfolio management for mixed holdings of cash and risky securities. The CAPM is credited to Sharpe (1964), Lintner (1965), Mossin (1966), and an unpublished paper by Treynor. The single-index model was developed by Sharpe (1964). Ross (1977) simplifies the derivation of the CAPM and clarifies theoretical issues concerning the effects of alternative assumptions concerning short sales and the risk-free asset in the CAPM. In particular, Ross shows that the efficiency of the market portfolio and the CAPM are equivalent; if the market portfolio is inefficient the CAPM will not hold.

The following are references to selected areas of portfolio optimization literature that are particularly interesting from the perspective of approximate methods:

1. *Criticism of portfolio analysis.* Reliability of portfolio estimates, see Frankfurter, Phillips, and Seagle (1971), Dickinson (1974), Barry (1974), Barry and Winkler (1976), Frankfurter and Phillips (1979), Jobson and Korkie (1980, 1981), Frost and Savarino (1988), Michaud (1998, 1989), Markowitz (1991), Haugen (1995, 1996, 1997, 1999a, and 1999b). Jarrow and Madan (1997) argue that a static, two-parameter criterion is too simple in sophisticated financial markets and seems vacuous to these authors; further, they argue that the concept of beta is stillborn.

2. *Portfolio optimization.* Martin (1955), Sharpe (1967) linear approximation to mean-variance analysis by diagonalizing the variance-covariance matrix, Sharpe (1971) LP algorithm which concentrates on systematic risk, Francis and Archer (1971) explore Martin's basis reduction method, Merton (1972) provides an analytic derivation of the efficient frontier under a variety of assumptions. The EGP algorithm and original sources can be found in Elton and Gruber (1995), and Kwan (1984) unifies alternative ranking techniques within EGP methods. Tarrazo (2000b) shows portfolios can always be optimized using SES methods. Fogler (1993) notes that combinatorial complexity in security selection may push analysts into traditional methods of security analysis. Lee (1990) shows that the length of investment horizon matters if asset prices do not follow a random walk.

3. *Short sales.* Lintner (1965), Renshaw (1977), and Markowitz (1983) are "must read" references for anyone interested in portfolio analysis. Levy and Sarnat (1984), Sharpe (1991) CAPM with and without negative holdings, Miller (1990, 1991) notes that nonlinearities are likely under alternative assumptions concerning short sales, Alexander (1993, 1995), and Tarrazo and Alves (1998b).

4. *Size of portfolios, limited diversification, and transaction costs.* Jacob (1974), Brennan (1975), Lease, Lewellen, and Schlarbaum (1976), Statman (1987), Wagner and Banks (1992), Genotte and Jung (1992), Newbould and Poon (1993), and Elton and Gruber (1977).

5. *Other.* Merton (1980) on estimating the rate of return on the market, Leibowitz and Henriksson (1989) confidence limit approach to manage downside portfolio risk, and Smith (1968). Bogle (1992) on investing in the S&P500 Vanguard Index Trust.

The best way to reformat, defragment, and reoptimize one's knowledge of portfolio analysis—as one maintains a hard-disk loaded with information—is this:

First, thoroughly study a comprehensive text such as Haugen (1997), Elton and Gruber (1995), or Tucker et al. (1994).

Second, select some particular aspect of portfolio analysis—I prefer short sales and the optimization itself—and use it to penetrate the thick ball of portfolio literature. The literature is overwhelming and one needs a strategy for "ingress" and "egress," to use astronomy terms.

Third, even investors who specialize in equity portfolios rarely think only in terms of equity. Fixed-income markets and derivatives are moving parts of the investment machinery and they all affect each other. Therefore, one must also keep a basic, fundamental, and updated knowledge of fixed income and derivatives. The former can be obtained from Fabozzi (1993), the later from either some chapters in Haugen (1997) or from specialized textbooks such as Chance (1995), Jarrow and Turnbull (1996), and Luskin (1987). Two textbooks on derivatives that emphasize pricing issues are Ritchken (1987) and Cox and Rubinstein (1985). Two books on derivatives that emphasize trading aspects are MacMillan (1986) and Bookstaber (1989).

Fourth, at some time our knowledge of the tools we use in portfolio analysis will need fine-tuning. In my personal library I have several sources I routinely consult. I have experienced that the most important doubts usually concern fundamental material, which has been superbly dealt with in accessible textbooks. Luenberger (1988) and Chiang (1984) are recommended references for classical optimization and mathematical programming, Freund (1992) and Spiegel (1992) for statistics, Judge et al. (1985) and Johnston (1984) for econometrics, and Lipschutz (1991, 1989) for linear algebra.

Fifth, portfolio analysis is simply an implementation of financial thinking. The same reasoning concerning risk management, hedging, and strategy-making in complex and sophisticated settings should be present in corporate finance and could be present in the financial management of our own households—a non-trivial issue, as you know. This means one can learn a great deal about portfolio management by learning about financial reasoning. In this context, Copeland and Weston (1988) provides the best vehicle to upgrade our financial knowledge and reasoning. Bodie and Merton (1998) is also recommended in this respect.

# EXTENSIONS
# TO APPROXIMATE
# EQUATIONS MODELING

When I started writing this book, I thought it would be a good idea to close it with a chapter on extensions to the approximate equations methods presented. I wanted to develop a base of cases for further application of approximate equations. For example, microeconomic models (consumer demand, production theory, market models, and general equilibrium) use simultaneous equations systems (SES) that perhaps can be improved upon using approximate ones (AES). However, after my exploration of approximate equations I have different types of extensions to present, which may strengthen the case for applying AES to other decision-making models.

This chapter builds on the following five ideas concerning the effects of taking imprecision into consideration when we develop a model.

One. In the social sciences the researcher imposes the model on "the world." We cannot say "If the system has a second order differential equation" as it may be said in other disciplines. We impose what we want depending on our theories, expertise, and preferences. This means, as it has been pointed out by Caws (1988), that most of what we think we know in the social sciences is built by ourselves. We work with representations, as was also observed by Kant and Schopenhauer.

Two. The AES bridge between classical (calculus and linear algebra) and operational (mathematical programming) methods is worth exploring again, because the differences between classical and programming methods may be lessened in the presence of imprecision (and other nonprobabilistic uncertainty).

Three. The AES bridge between SES and alternative dynamic specifications is also worth exploring for the same reason. Moreover, the assessment of the usefulness of modeling issues such as nonlinearity and stochastic specifications changes when the starting model is an AES. We may not know enough to discriminate among the very precise dynamic alternatives commonly available.

Four. The search for significant structure, especially given ideas one, two, and three above, is the most important activity in modeling. Our presumption is that finding some significant structure, which can later be modeled as an approximate equation system, is more important in practical terms than mere technical sophistication in the modeling tools.

Five. The choice between words and numbers is the most important decision to be made by researchers in social sciences, including those in economics and finance. Approximate equations models add error to exact systems, which also adds a "grain of salt" to the theories themselves. This will make numbers less precise and, perhaps, takes them closer to words. The same reasons that compel us to use AES, therefore, point out to exploring qualitative and words-based methodologies in economics and finance.

The structure of this chapter follows the flow of the ideas above. The first part is dedicated to exploring the link between AES, SES, and programming methods. The second part studies dynamic modeling from an approximate perspective. These two sections indicate that in some problems AES may be all the optimization and modeling tools the researcher may need. The reader may notice that ideas two and three expressed above could require rather involved scholarly studies, if not entire books, to fully explore. This chapter will not try to develop those ideas, but simply to outline their rationale. The overall intention of this chapter is to leave the reader with the perception that there is something in approximate equations that warrants further attention, and that this is a practical value that some of the conventional, more exact, mathematical apparatuses fail to exhibit. This chapter also engages in a critical review of fundamental methods in modeling. Uncritical expositions—mere review of basic material—abound and are only helpful to uninformed readers. Critical reviews, however, aim at clarifying the importance of what we usually take as given. They are always informative and generally contribute to better treatment of practical problems.

The third part of this chapter will provide observations on the relationship between approximate equations systems and other modeling alternatives.

The concluding remarks in this chapter will serve as the epilog for this book.

## 1. MODEL SELECTION AND CHANGE

The starting point in this book was the simple simultaneous equation model $A x = b$. It has been presented throughout as the fundamental modeling specification because it presupposes (and requires) a very detailed knowledge of the problem at hand and very precise definition and measurement of the variables involved. What makes this model so important is also what limits it: it requires a level of precision we usually do not have in practice. In addition to its precision requirements, Chapter 2 noted that the SES specification is also limited to handle change, which diminishes modeling success. One of the themes of this book is that approximate equations can preserve some of the strengths of SES modeling—classification of variables into exogenous and endogenous, richness of structure, equilibrium interpretations between controllable and external variables, and so on—and also address some of their limitations. This chapter, like Chapter 2, starts by studying the effects of change on a simultaneous equations system such as $A x = b$, or more generally $F(x, b) = 0$, where $F(x)$ represents the structure of the problem.

First, note that optimization is a variety of equilibrium analysis. Optimization, however, generally includes constraints, which makes the problem more realistic. Optimization problems can be solved with classical methods (calculus-based SES systems, calculus of variations, and so on), or with operational methods (mathematical programming). Mathematical programming problems and classical problems have much more in common than is generally accepted. For example, a typical problem in optimization uses a twice differentiable function $F(x)$, subject to linear constraints in the form of one or more functions, $G(x)$. To optimize is to find a set of values for key variables ($x_i$'s) that provide the best values for $F(x)$, subject to $G(x)$. When $F(x)$ is quadratic, this model leads to a simultaneous equation system using Lagrange's method.

When we handle optimization problems the effects of change are assessed with what is called postoptimal analysis. When we study equilibrium problems, the effects of change are studied via comparative static analysis and perturbation theory, which can be understood as varieties of postoptimal analysis. As we saw in Chapter 2, perturbation theory (PT) studies changes around the equilibrium point $x^*$. For small changes in b and A ($\Delta b$, $\Delta A$), the corresponding $\Delta x$ can be approximated by:

$$(1) \qquad \Delta x = (A + \Delta A)^{-1} \Delta b + (A + \Delta A)^{-1} x^*$$

or by $\Delta x = (A + \Delta A)^{-1} x^*$, when $\Delta b=0$ and second-order effects are neglected. PT is a form of comparative statics (CS). However, CS usually aims at assessing effects of change in individual coefficients, $x_i$, brought about by a change in the structure of the system (A), independent of initial conditions ($x^*$): $dx_i/da_i$. It is critical that we make it very clear that optimal analysis rests on compensating changes: a given change in a closed system in equilibrium brings about other changes that must be considered for the analysis to make any sense. Note that A expresses first order optimal conditions for constrained optimization such as the following:

$$(2) \qquad L = F(x_1, ..., x_k) - \sum_m \lambda_m (G_m(a_{1m} x_i, ...., a_{km} x_k) - G_{0m})$$

$$(3) \quad \begin{vmatrix} f_{11} & ... & f_{1k} & g_{11} & ... & g_{1m} \\ ... & ... & ... & ... & ... & ... \\ f_{k1} & ... & f_{kk} & g_{1k} & ... & g_{km} \\ g_{11} & ... & g_{1k} & 0 & ... & 0 \\ g_{1m} & ... & g_{km} & 0 & ... & 0 \end{vmatrix} \begin{vmatrix} x_1 \\ ... \\ x_k \\ \lambda_1 \\ \lambda_m \end{vmatrix} = \begin{vmatrix} 0 \\ 0 \\ 0 \\ G_{01} \\ G_{0m} \end{vmatrix}$$

$$\qquad\qquad A \qquad\qquad\qquad\qquad x \quad = \quad b$$

The endogenous-variable Jacobian (first-order conditions), which is equal to the bordered Hessian (H), A in equation (3), is obtained via implicit function theorem and indicates that $(dx_i/da_i)_m = {}^mH_{ii}/{}^mH < 0$.

Up to this point, it is evident that there are three elements in A x = b, and that if x* = A⁻¹ b exists, then we can study change in this model by a number of simple procedures that involve each of those elements: x, A, and b. These procedures do not exhaust the ways in which we can develop models when change affects a SES specification. Exhibit 6.1 presents a classification of those ways.

In addition to changes in b, and joint changes in b and A, we can endow our model with dynamic properties, as the third part of this chapter will study. It is important to remember, however, that as long as we hold on to A x = b, we limit the number of ways in which the model can be made dynamic. The optimization setup, x* = A⁻¹ b, includes much information and foretells what we will find after the optimization: existence of equilibrium, its stability, and also the seeds of its dynamic behavior—what Samuelson (1983) referred to as the "correspondence principle." Therefore, it appears that we will have to abandon the A x = b model and develop a totally different one. This was also the perception when the SES model was abandoned in favor of a programming one; and in a very similar way, when we think we are doing something different we find we are simply handling a different SES model. The third part of this chapter will illustrate this.

**Exhibit 6.1**
**Studying Change in the SES Model**

1. Changes in b
2. Changes in b & A
3. Dynamic models
4. Model selection
   a) Precision
   b) Nested alternatives
   c) Non-nested alternatives:
      c.1 Missing variables and equations
      c.2 Functional form
      c.3 Alternative dynamic specifications

Item 4 in Exhibit 6.1 has the heading, "Model selection," and includes three options. The first makes reference to stochastic versus nonstochastic specifications. Our objections to probabilistic systems have been made in different parts of this manual on application of approximate equations. Probabilistic modeling, in many cases, represents a smokescreen behind which (1) we hide what we do not know about a model, and (2) force the model to exhibit properties (limiting distributions, convergence, ergodic tendencies, and so on) that we ourselves predetermine. In practically every case we have seen in economics and finance, simulation analysis has been used in this way. To the contrary, approximate equations place before our eyes the uncertainty box the model carries within itself. This is important when the model is used to support real decision-making.

When studying model selection, the distinction between nested and non-nested alternatives is useful. Two alternative models are nested when they share the same fundamental structure; for example, adding or eliminating one restriction in the model 2–3 above. Non-nested alternatives, however, represent completely different structures and, therefore, can be very hard to compare. This classification and much of the discussion on modeling in this chapter is highly influenced by our formation in econometrics, see Judge et al. (1985).

Finally, a model is a characterization of a problem and has the following three elements: mathematical objects, theories, and data. In this chapter we do not deal with theories, which in some cases determine the mathematical objects to be used. Unfortunately, in economics and finance it happens many times that the same theory can be modeled in a number of ways.

## 2. AES AND MATHEMATICAL PROGRAMMING

The previous section noted how programming problems and classical methods amount to solving a simultaneous equation system as in, for example, solving a quadratic function subject to linear constraints. First-order conditions in the optimization of the aforementioned programming problem generated a set of simultaneous equations. This section explores the relationship between linear programming and simultaneous equations problems. It first reviews the relationship in exact terms, and then from an approximate viewpoint.

We learned about SES when we were about 12 years old. It was five years later that we became first acquainted with LP methods. We adopted the usual textbook line that LP methods are superior to SES because they can handle inequality constraints, and because they could solve a problem where the number of variables exceeds the number of equations. This is, however, a perception we have had ample opportunity to refine over the years, especially when we came across approximate equation systems.

Let us recall the expression for a SES system of order k:

$$
\begin{array}{ccc}
(4) & A & x & = & b \\
& (k \times k) & (k \times 1) & & (k \times 1)
\end{array}
$$

Let us compare it to the typical programming problem:

$$
\begin{array}{lll}
(5) & \text{Optimize} & p & x \\
& & (1 \times k) & (k \times 1) \\
& \text{subject to} & A & x & \leq & b \\
& & (c \times d) & (d \times 1) & & (d \times 1)
\end{array}
$$

where $d = c \diamond k$, and $x_j \geq 0$ for some $j = 1, \ldots, q$.

At first sight these systems look very different. However, they are the same if:

1.  $d = k$
2.  non-negativity restrictions are met, and
3.  the solution is found for the best (least restrictive) case scenario of restrictions met with the equality sign.
4.  In this case, the set of restrictions becomes a regular simultaneous equation system and the objective function is totally accessory. Let us study each of these events, that is dimensionality, inequalities, and sign restrictions.

With respect to dimensionality, a full rank for A means that each variable supplies an equation that brings new information into the problem. When A does not have a full rank (linear dependency, missing equations, and so on), solution points cannot be found for all $x_i$, but only for those that brought their valid equations to the problem. There can also be more equations than variables. In that case, $d > c$, and the system is overdetermined. In each of these cases, $d \neq k$, classical methods run into trouble, but the simplex algorithm of linear programming comes to the rescue and finds a solution by using the information provided by the objective function. When $d > c$, the simplex method will select those constraints that effectively bind the objective function variables at their optimum value. When $d < c$, it means there are less restrictions than variables and the system represented by the constraints would have many solutions. However, the objective function is there to reveal which variables are important or valuable in the resolution and the obedient simplex algorithm will select those from the system represented by the constraints in the most efficient and in this case comprehensive manner: "d" steps. We could say that the objective function is less useful as an indicator as the number of effective constraints approaches the number of variables.

What about the inequality restrictions? There are two distinguishable cases. When the inequalities do apply, the simplex method uses special "slack" variables to create equality constraints. These slack variables do not have a zero value at the end of the optimization. When the inequalities are met with equality signs, the slack variables will have a zero value at the end of the algorithm. Strict inequality restrictions ($<$ ) are a problem for simultaneous equation systems. If the first equation represents a strict inequality in a two-variable SES system, there would be $a_{11} x_1 + a_{12} x_2 < b_1$, which could be rewritten as $a_{11} x_1 + a_{12} x_2 = b_1 + c_1$, $c_1 > 0$. The system would be

$$(6) \qquad \begin{aligned} a_{11} x_1 + a_{12} x_2 &= b_1 + c_1 \\ a_{21} x_1 + a_{22} x2 &= b_2 \end{aligned}$$

However, the system would be underdetermined if the value for $c_1$ is unknown. If A, b, and c1 are constants, we could obtain a set of solutions for $x_1$

and $x_2$ that would depend on the value of the unspecified $c_1$. The LP model for this case would be

(7)           Max     $p_1 x_1 + p_2 x_2$
              s.t.    $a_{11} x_1 + a_{12} x_2 < b_1$
                      $a_{21} x_1 + a_{22} x_2 = b_2$

which is easily solved.

Remember, however, that with nonstrict inequalities ($\leq$) the best case scenario coincides with that of SES. Note now that there are many ways in which SES methods and LP methods are linked. One of them is, as this discussion shows, the theory of inequalities. Another one concerns the area of integer programming, where LP methods can be related to those of finding the solution to integer simultaneous equations systems (diophantine analysis), see, for example, Nemhauser and Wolsey (1988).

What about sign restrictions for optimal variables? The signs of the optimal variables are extremely important. For example, in the portfolio optimization case, a positive optimal weight means the stock is to be purchased (held long), while a negative optimal weight means the stock should be sold short. Signs are critical in programming because they determine feasible solutions. The set-up of simplex algorithm in linear programming precludes negative solutions. Nonlinear programming methods often apply an ad hoc logical rule of eliminating those negative variables from the basis. For example, the exclusion rule applied by many nonlinear programming algorithms is simply to remove negative variables from the basis because they are unfeasible. For example, Wolfe's method for nonlinear programming applies the exclusion rule at the constraint (A x = b) level and, therefore, is an operation on the SES system, see Foulds (1981, p. 359). (Some material concerning sign restrictions may be found in Boot (1964, p 31.), and in Franklin (1980, p. 57). Franklin discusses Fredholm's alternative of linear algebra as an introduction to a well-known theorem by Farkas, which leads to the separating hyperplane theorem, duality theorem, and Khun-Tucker conditions of both linear and nonlinear programming.)

The previous chapter (see Tarrazo, 2000b for a more formal treatment), showed how a basis reduction method yields the optimal, nonnegative, solution without ever using mathematical programming.

In sum, we can say that LP and its simplex algorithm is a "Jack of all trades" that either selects the best variables when there are too many of them, or the effective constraints when some of them may be accessory. In addition, the simplex method handles inequality constraints in the restrictions. The simplex algorithm is most useful when information is most scarce. For example, the duration-based immunization problem tries to find a bond portfolio that will be unaffected by certain changes in interest rates. It amounts to maximizing the yield of the portfolio, expressed as the average of individual bond yields, subject to two restrictions: one concerning durations and the other the full wealth constraint that the summation of optimal weights adds up to one. The simplex algorithm finds the best two bonds out of a set of two, three, 10, 40, 100 bonds, and so on, and does so in two steps. This is important: the dimension of the solution space in a linear programming problem is equal to the number of effective constraints in the solution; we find two optimal bonds, in two steps, because we have two active or effective constraints. As noted earlier, the objective function is less useful as an indicator as the number of effective constraints approaches the number of variables.

The main idea driving our review of programming methods is that there is a strong relationship between SES and those methods in exact modeling. The following observations stress this idea:

1. Every point in $\Re^k$ can be written as the solution to a SES system; therefore, the solution $x^* = \{x_1^*, x_2^*, \ldots, x_k^*\}$ found in a programming problem is the solution to a SES system as well. In the case of a LP problem, the relevant SES comes from the structure of the restrictions and its objective function, which may play a more or less important role. In the case of a QP problem, the SES structure is given by the Lagrangian optimization as we saw in portfolio optimization.
2. The simplex method uses a type of special variable—"slack" variables—to turn inequalities into equalities and handle restrictions. Lagrange's method substitutes restrictions for special variables—multipliers.
3. The solution to an AES actually implies application of mathematical programming algorithms. Therefore, if the AES represents the actual structure of the problem, it may also include programming solutions.

In an approximate context, the relationship between classical (SES) and operational (programming) methods turns into a solid bridge. Again, the following observations should add weight to our presumption:

1. The objective function may be regarded as a missing equation, which somehow may be approximated within an AES context.
2. The slack variables in a programming problem may be regarded as errors. This is, in fact, how programming systems are solved.
3. When a programming problem exhibits fewer constraints than a SES equation ($d < k$), we can view it as a case of missing variables, which again brings to mind an approximate equations system, as noted in Chapter 2.

The challenge now is to illustrate with the help of examples what has been noted concerning the relationship between SES and programming methods, especially in an approximate context, and to do so with an eye for practical applications. For the exact case, let's illustrate that relationship with an example from Chiang (1984, p 689), which is presented in Exhibit 6.2. In this simple case, the objective function represents a missing equation from the set of restrictions. When the objective function is "plugged" into the set of constraints, we obtain a SES system to which we can apply approximate procedures. These approximate methods can, in turn, inform us about potential ranges of uncertainty for the variables involved if there were uncertainty concerning the model itself.

**Exhibit 6.2**
**SES and Programming Methods**

Problem 1

Maximize $p = 3 x_1 + 4 x_2$
s.t.

$$\begin{vmatrix} 1 & 1 & 3 \\ 2 & 4 & 1 \end{vmatrix} \quad \begin{vmatrix} x_1 \\ x_2 \\ x_3 \end{vmatrix} \quad \leq \quad \begin{vmatrix} 12 \\ 42 \end{vmatrix}$$

$x_i^* \geq 0$, for $i = 1,2,3$

$x^* = \{ 3, 9, 0 \}$, $p = 45$, $r_1 = 12$, $r_2 = 42$

Problem 2

SES system:

$$\begin{vmatrix} 3 & 4 & 3 \\ 1 & 1 & 3 \\ 2 & 4 & 1 \end{vmatrix} \quad \begin{vmatrix} x_1 \\ x_2 \\ x_3 \end{vmatrix} \quad = \quad \begin{vmatrix} 45 \\ 12 \\ 42 \end{vmatrix}$$

$x^* = \{ 3, 9, 0 \}$

Ranges for the approximate version of this system:

| | Error = | 0.12 | |
|---|---|---|---|
| | Max | Centered | Min |
| $x_1$ | 12.75 | 3 | -18 |
| $x_2$ | 19.5 | 9 | 4.125 |
| $x_3$ | 5.25 | 0 | -2.4375 |

The usefulness of applying this technique to a LP problem, therefore, depends on the problem itself and, above all, the amount of information and exact knowledge we have. There are problems in programming with a lot of structure (d close to k, the diet problem for example), or with very little structure, Chiang (1984, pp. 652, 657, respectively). Also keep in mind that it can be very difficult to study change in programming models, see for example Gal (1979), and that dynamic operational methods tend to be both cumbersome and limited in their applicability. For example, in dynamic programming one can only have an exogenous variable changing.

As noted earlier, the case of portfolio optimization, studied in the previous chapter, illustrates well the relationship between SES and programming methods in an approximate context. As the reader may recall, the no short-sales solution is actually the LP-SES solution. Therefore, the AES solution encompasses the LP solution and something else. Given the imprecision of the system (lack of information) even some positive weights could be negative! One could never be aware of this important fact if programming solutions are used. The nonnegative solution can also be obtained by adding auxiliary constraints disallowing the type of financing generated by short-selling, see Tarrazo (2000b).

Chebyshev's approximations are another example of the relationship between SES and programming methods in an approximate context (see, Franklin, 1980, p. 8). What is known as "multiobjective programming" amounts to using several objective functions, which brings SES and programming methods closer together (see Cohon, 1978, especially page 72).

I am not advocating against using programming methods, nor am I saying that they are not useful. The role of programming methods is very important in some problems. Still the relationship SES-LP remains. For example, in game theory every matrix game can be reduced to a linear optimization problem and every linear optimization problem can be reduced to a matrix game. In any case, there is a link between classical (SES) and operational methods (programming), which turns out to be even more important in an approximate context, and this link can help in decision-making models, which in economics and finance are always approximate.

### 3. AES, OPTIMIZATION, AND DYNAMICS

In a way, modeling SES in an approximate form represents an alternative to some techniques for postoptimal analysis. AES calculate all possible ranges for the variables involved in the problem and, therefore, exhaust the searches in a solution space larger than that considered by the solution to $A x = b$. It is time to consider dynamic issues. Our main idea in doing so, as noted in the beginning of this chapter, is to indicate that the link between the (static) $A x = b$ and other dynamic specifications (remember Samuelson's "correspondence principle") becomes more important in the case of AES. The intuition is that if our knowledge of the problem allows us to work with a static simultaneous equation system, but we do not have information concerning its dynamics other than it is not an explosive system, the approximate version of the SES specification may serve to encapsulate plausible dynamics. (An "explosive" system is one that moves irretrievably away from its initial situation.)

Dynamic treatments of a model can be motivated in a number of ways. For example, Samuelson (1983) argues that comparative statics and other postoptimal analyses are not really useful if nothing is said about the dynamic properties of the variables and the system under analysis. We can say that if there is a change in $A_0 \, x_0 = b_0$ the new equilibrium should be $x_1$*, but how do we know that the system will transition from the original values to the new ones without making an explicit hypothesis regarding its dynamic behavior?

Another way to motivate dynamic analysis of a given model is to think that users of the model work, after all, with points of reference. For example, when we invest in a mutual fund we may look at the value of the S&P500 Index (I) in order to assess and evaluate the change in our own portfolio (P); that is, we could link our portfolio to the index as in $P = a + b\,I$. We could also study the temporal evolution of our portfolio independently of any other, as when we posit $P_t = a + b\,P_{t-1}$. Note that in this last formulation, we need the help of time to assess change. The last formulation is so useful that we often employ a new variable, portfolio returns $(r_t)$, which are defined with reference to a time frame. The equation for i-period returns can be written as $r_t = (p_t - p_{t-1})/p_{t-1}$ in the discrete time case, and $r_t = \ln\,(p_t/p_{t-1})$ for the continuous case, where "ln" represents the natural logarithm. However, note that we could also posit $P(t) = f(b\,I(t))$, where both variables are time dependent and the relationship does not need to be linear.

Independent of our motivation, introducing time considerations to the analysis makes it realistic and, it is hoped, more practical.

This section will first outline dynamic specifications and then offer some critical comments.

Technically, there can be what are called "autonomous" systems, which are those that run their course independent of time. But when researchers do not include time in their models it is due more to lack of knowledge of the problem than to specifically knowing that time does not play a role, or that it will not have any "reference" value.

Exhibit 6.3 shows the many options available to develop dynamic models. The first option is to choose a differential or a difference equation. "Differential" means that the time and variable increments are infinitesimal, while "differences" express larger changes (increments). This choice, therefore, should be determined by the type of available data, but there are other considerations concerning formulation of solutions and time behavior, as we will note. Very loosely stated, differential equations are studied in "analysis"—the part of mathematics that studies objects such as differential and integral calculus, functions, and variables—while difference equations seem closer to algebra and operational methods. (When working with differential equations one can sometimes interpret first derivatives with respect to time as the past of the function and the integral may be taken to represent the future. With differential changes we use integrals "$f$," with differences we use summations "$\Sigma$.")

Following the order of Exhibit 6.3, the next choice would concern existence of a variable independent term. We can make it equal to zero (homogeneous case), or equal to a constant ("b" in the exhibit), or we can make it time dependent ($=b(t)$). This term, as the reader may have noted, is similar to b in our original SES and, in a similar way, indicates the existence of a time-independent (b), or a time-dependent ($b = b(t)$) equilibrium.

**Exhibit 6.3**
**Techniques for Dynamic Modeling**

|  |  | Differential | Difference |
|---|---|---|---|
| 1. | Independent term | | |
|  | a) Equals zero. | $dy/dt + a\,y = 0$ | $y_{t+1} + a\,y_t = 0$ |
|  | Homogeneous. | $dy/dt + a\,y = b$ | $y_{t+1} + a\,y_t = b$ |
|  | b) Nonhomogeneous: | | |
|  | b is not equal to | $dy/dt + a\,y = b(t)$ | $y_{t+1} + a\,y_t = b_t$ |
|  | zero (Constant | | |
|  | versus variable | | |
|  | term, e.g., $b = b(t)$). | | |
| 2. | First or higher order | $y'' + a1\,y' + a2\,y = b$ | $y_{t+2} + a1\,y_{t+1} +$ |
|  | (The example given is a | | $+ a2\,y_t = b$ |
|  | nonhomogeneous, | | |
|  | second-order process) | | |
| 3. | Coefficients ai | | |
|  | a) Constant | Any of the above | Any of the above |
|  | b) Variable | $y' + u(t)\,y = b$ | $y_{t+1} + at\,y_t = b$ |
|  | | | $a_t = \alpha0 + \alpha1\,a_{t-1}$ |
| 4. | Linearity | | |
|  | a) Linear | Any of the above | Any of the above |
|  | b) Nonlinear | Products of y' and | Products of y' |
|  | | higher orders or | and higher orders |
|  | | powers for the | or powers for the |
|  | | differentials | differentials |
| 5. | a) Single equation | Any of the above | Any of the above |
|  | b) Simultaneous | See text | See text |
|  | equations systems | | |

Notes:

For differential equations: $dy/dt = y'$ , $d_2y/dt^2 = y''$ , etc.
For difference equations: $\Delta y = y_{t+1} - y_t$, or $\Delta y = y_t - y_{t-1}$

Another choice concerns the order of the specification. We can only increase the order of a specification if we have correspondingly more accurate knowledge of the dynamics of the problem. In other words, when we know little about a process we can only lay out very simple (low order) dynamics. More will be discussed about this later on.

Another choice concerns the specification of the coefficients of the equation. It is possible to posit time-dependent coefficients in a dynamic model, and computer programs exist that would "try out" different dynamic specifications automatically. The problem is whether we know enough about the problem to justify the "trying out" activity, which has become a problem in econometrics— see, for example, Leamer's criticism of specification searches in Judge et al. (1985, several chapters). Dynamic modeling requires very precise knowledge of the problem being considered.

Another difficult choice concerns linearity. As in the case of the order of the process, variable coefficients, and time-dependent terms, nonlinear specifications require very detailed knowledge of the problem under consideration. Nonlinear options require better knowledge of the problem and complicate the solution process. With linear systems, the solution can be found by splitting the problem and finding and adding partial solutions (superposition theorem), which cannot be done with nonlinear specifications. There is something else. Linear systems cannot have self-adjusting properties, which complex systems usually exhibit. Research on chaos theory has shown that very simple models (a couple of equations with a couple of variables or even a second-order equation with a single variable) can exhibit very sophisticated dynamic behavior. This behavior, if it is nonlinear, may help to explain the behavior of apparently very complex systems (See for example Strogatz, 1994; Drazin, 1992; and the nontechnical account of Gleick, 1987). Linearity is something we should not discard beforehand, but it is also something extremely difficult to incorporate into a dynamic model, especially into a stochastic one. Samuelson (1983), and Chiang (1984), whose notation we are adopting in this part of the chapter, contain many examples of dynamic modeling in economics. Begg (1982) and Barro (1990) contain many macroeconomic examples. In economics and finance we can say that we do have some models, but we have no idea of whether they work or not. We are focusing on nonprobabilistic models because our approximate models handle uncertainty in a different way. Assumptions concerning stochastic patterns end up dominating dynamic behavior in stochastically dynamic models, which allow the researcher to obtain whatever he or she sets up to obtain, at least in macroeconomics. Furthermore, most of the time errors are "tamed" to be well behaved . . . why then do we add them?

The last choice, and all these choices are hard to make, concerns dynamic single versus dynamic simultaneous equations specifications (SES-D). A priori, these models seem to represent the last echelon of complexity in modeling and they are so. However, what we know in one equation may help in another, and what we ignore in one equation may be bounded by what we do know in another. Therefore, there is some hope that SES-D systems may enable at least a better understanding of the problems we are facing. As this section shows later, there is also something about them that makes them particularly attractive and brings them closer to AES.

Let us start with the simplest possible case: homogenous, first-order equations. A first order differential equation of this type can be expressed as $dy/dt + a\,y = 0$, or $dy/dt = a\,y$. This means that $1/y\;dy/dt = a$, a constant. This means that the rate of growth (proportional change with respect to time) is constant. The reader has handled this type of formulation because, for example, the growth of funds on an interest-bearing account can be modeled in this way. The constant, a, would represent the rate of interest. The general solution for such an equation is $y(t) = A\, e^{-at} = A \exp\{ - a\, t \}$, if the initial amount in the account is known to be $y(0)$, the definite solution or value of the function at a particular point in time is $y(t) = y(0)\, e^{-at}$. (Taking logarithms and differentiating with respect to time gives rates of growth. For example, taking logarithms in $y(t) = y(0)\, e^{-at}$ and letting $t = 1$ provides the equation for rt shown earlier.) The case of a difference equation (first-order, homogeneous) is also easy to understand. The general solution to an equation such as $y_{t+1} + a\, y_t = 0$ is $y_t = A\, (-a)^t$. Time behavior in difference equations depends on the absolute value of the root $= -a$, while in differential equations it depends on the sign of the exponent a.

Simple first-order, nonhomogeneous cases of differential and difference equations show that the general solution (yg) is the sum of a particular solution (yp) and the complementary function (yp), that is $yg = yp + yc$. For the differential equation case, we would have

(8)   Function:                      $dy/dt + a\, y = b$
       Particular solution:           $yp = b/a$ ; (implies $dy/dt = 0$)
       Complementary function:  $yc = [y0 - b/a]\, e^{-at}$
       General solution:              $yg = yp + yc = c/a + [y0 - c/a]\, e^{-at}$

For the difference equation case:

(9)   Function:                      $y_{t+1} + a\, y_t = b$
       Particular solution:           $yp = b/(1 + a)$ ; (implies $y_{t+1} = y_t$, and $a \neq 1$).
       Complementary function:  $yc = A\, (-a)^t$
       General solution:              $yg = yp + yc = [b/(1 + a)] + A\, (-a)^t$

There is one technique that can be employed to find solutions for the case $a = -1$.

These two cases show the role of the complementary function. It activates itself when the variable is out of its equilibrium, or time-independent, solution yp. If we let $t = 0$, we can check that the general solution provides the particular one. When t is increasing (t cannot decrease) the value of the complementary function will vanish, stay stable, or decrease, depending on the (absolute) value of the root a (difference case), or the sign of the rate of growth (differential case). Therefore, when $a > 0$, as in the differential case, the term $e^{-at}$ will tend toward zero as t grows. The difference equation case is more complex. Let the root be $r = (-a)$, and we can distinguish seven dynamic regions:

| | | |
|---|---|---|
| I. | $r > 1$ | Explosive, divergent |
| II. | $r = 1$ | Unchanged at a fixed distance from yp |
| III. | $0 < r < 1$ | Amortiguated, convergent |

| IV. | r = 0 | Unchanged |
| V. | −1 < r < 0 | Convergent with oscillations |
| VI. | r = -1 | At a fixed oscillating distance from yp |
| VII. | r < -1 | Explosive with oscillations. |

The multiplicative constant produces a "blow up" or "pare down" effect on the complementary function, which is also known as a "scale" effect (see Chiang, 1984 for further detail).

Solving higher-order difference and differential equation systems amounts to finding multiple roots for the dynamic system. A second-order system (see the first half of Exhibit 6.4) would provide a quadratic equation to be solved. This equation is called a characteristic equation. The dynamic behavior of the system depends upon the roots of this equation, which can be: (1) real and different roots, (2) equal real roots, or (3) complex roots. Dynamic equilibrium depends, as in the first order case, on the complementary function yc $\rightarrow$ 0, when t $\rightarrow$ ∞ . In the case of differential equations, if the roots are both negative, the system will be dynamically stable no matter what the initial condition is. In the difference equation case, the requirement is that the absolute value of every root must be less than one. The study of dynamic behavior in the complex-roots case must be handled with transcendental functions (sine, cosine, and so on). Note that the solution for the complementary functions for linear systems, independently of the order, can be conveniently expressed as a linear combination of variables.

Exhibit 6.4 presents simultaneous equations: a nonhomogeneous, differential system, and a difference one, respectively. It is well known how to find solutions to these linear systems. What makes modeling with SES-D difficult are the following elements, which must all be taken into account: (1) classify variables into exogenous and endogenous, if needed; (2) specify the dynamic behavior of exogenous variables; and (3) specify the dynamic behavior of endogenous variables, which depends upon feedback (autoregressive) effects and interactions with both exogenous and other endogenous variables.

We must now outline two relationships. The first is that between the static SES, A x = b, and its dynamic counterparts (SES-D). The second is between these exact models and approximate versions of a SES (AES).

With respect to the first relationship, note first that the particular or equilibrium solutions of dynamic specifications are actually the time-independent or unchanging $x^* = A^{-1} b$. This is obvious in the case of differential equations, and can also be easily seen to be the case with difference equations. For example, recalling perturbation analysis, given $\Delta x = (A + \Delta A)^{-1} x^*$, if we make $\Delta A = 0$, then $\Delta x = A^{-1} x^*$, which lets us attach dynamic meaning to the static SES model. Jacobi developed an iterative method for solving simultaneous equations systems, which also stresses the relationship between static and dynamic versions of simultaneous equations systems. It is interesting to briefly review his ingenious method.

**Exhibit 6.4**
**Further Examples of Dynamic Systems**

---

1. Second-degree differential equation (*)

Differential function: $\quad y'' + a1\, y' + a2\, y = b$
Particular solution: $\quad yp = b/a2$
Complementary function (**): $\quad yc = A1\, e^{r1\,t} + A2\, e^{r2\,t}$

2. Second-degree difference equation (*)

Differential function: $\quad y_{t+2} + a1\, y_{t+1} + + a2\, y_t = b$
Particular solution: $\quad y = b/(1 + a1 + a2),\ \text{for } (a1 + a2) \neq -1$
Complementary function (**): $\quad yc = A1\, r_1{}^t + A2\, r_2{}^t$

3. Simultaneous differential equations system (*)

Model:
$$\begin{vmatrix} 1 & 2 \\ 0 & 1 \end{vmatrix} \begin{vmatrix} x_1' \\ x_2' \end{vmatrix} + \begin{vmatrix} 2 & 5 \\ 1 & 4 \end{vmatrix} \begin{vmatrix} x_1 \\ x_2 \end{vmatrix} = \begin{vmatrix} 77 \\ 61 \end{vmatrix}$$

$$\quad\quad J \quad\quad\quad u \quad\quad\quad\quad M \quad\quad\quad v \quad\quad g$$

Particular solution: $\quad xp = M^{-1}\, g$

Complementary function (**):
$$\begin{vmatrix} x_1c \\ x_2c \end{vmatrix} = \begin{vmatrix} \sum m_i\, e^{ri\,t} \\ \sum n_i\, e^{ri\,t} \end{vmatrix}$$

4. Simultaneous difference equations system (*)

Model:
$$\begin{vmatrix} 1 & 0 \\ 0 & 1 \end{vmatrix} \begin{vmatrix} x_{1t+1} \\ x_{2t+1} \end{vmatrix} + \begin{vmatrix} 6 & 9 \\ -1 & 0 \end{vmatrix} \begin{vmatrix} x_{1t} \\ x_{2t} \end{vmatrix} = \begin{vmatrix} 4 \\ 0 \end{vmatrix}$$

$$\quad\quad I \quad\quad\quad u \quad\quad\quad\quad K \quad\quad\quad v \quad\quad d$$

Particular solution: $\quad xp = (I + K)^{-1}\, d$

Complementary function (**):
$$\begin{vmatrix} x_1c \\ x_2c \end{vmatrix} = \begin{vmatrix} \sum k_i\, A_i\, ri^t \\ \sum A_i\, r_i{}^t \end{vmatrix}$$

Notes:

(*) The general solution is always of the form: $yg = yp + yc$.
(**) Distinct, real roots case.

Given

$$
(10) \quad \begin{vmatrix} 1 & -2 \\ 1 & 2 \end{vmatrix} \begin{vmatrix} x_1 \\ x_2 \end{vmatrix} = \begin{vmatrix} -4 \\ 8 \end{vmatrix}
$$

one can single out one unknown per equation, $x_1 = -4 + 2\,x_2$, and $x_2 = (8/2) - (1/2)\,x_1$, and rewrite the system as

$$
(11) \quad \begin{vmatrix} x_1 new \\ x_2 new \end{vmatrix} = \begin{vmatrix} -4 \\ 8/2 \end{vmatrix} + \begin{vmatrix} 0 & 2 \\ -1/2 & 0 \end{vmatrix} \begin{vmatrix} x_1 old \\ x_2 old \end{vmatrix}
$$

The procedure feeds values for xi's and recycles the new values until the xi do not change anymore. Now note how time naturally slips into the system; some time elapses in each iteration, which lets us write xit+1 = xinew, and xit = xiold. Then, the system can be rewritten as

$$
(12) \quad \begin{vmatrix} 1 & 0 \\ 0 & 1 \end{vmatrix} \begin{vmatrix} x1_{t+1} \\ x2_{t+1} \end{vmatrix} - \begin{vmatrix} 0 & 2 \\ -1/2 & 0 \end{vmatrix} \begin{vmatrix} x_{1t} \\ x_{2t} \end{vmatrix} = \begin{vmatrix} -4 \\ 8/2 \end{vmatrix}
$$

$$
I \qquad u \quad - \quad K \qquad v \quad = \quad d
$$

Jacobi's ingenious strategy was to set up the system as a first-order dynamic one and let the dynamics of the system find the equilibrium values for x. The particular solution for this example, which is one of the AES employed in Exhibit 2.1, would be xp = (I + K)-1 d; then x* = {$x_1$*, $x_2$*} = {2, 3} as found in that chapter by doing x* = $A^{-1}$ b in equation (10). This example, however, is not well suited to study Jacobi's procedure, which works best with sparse and diagonal dominant A matrices. Our matrix A is not diagonally dominant: the absolute value of the sum of off-diagonal elements for a given row is not smaller than the absolute value of the diagonal ($a_{ii}$) element. Furthermore, the matrix K, which governs the evolution of the iterations, has roots of $r_1 = 1$ and $r_2 = -1$, which are equal. This means that starting the iteration from any value other than [2, 3], the optimal solution, will not produce convergence to those values. The Jacobi iterations are not helpful, which we already anticipated by observing diagonal dominance. Note that the system (10) is stable, with distinct real roots, or eigenvalues, of $\lambda_1 = -0.0705$ and $\lambda_2 = 0.0042$. The dominant eigenvalue is less than one in absolute value; the system is stable.

The technique of linearization employed in one of the macroeconomic models of Chapter 3 also highlights the relationship between static and dynamic SES. The fact that we can always transform an n-order differential equation system into a system of n first-order equations (see Chiang, 1984, p. 607) also stresses that relationship.

Second, the same characteristic equation is shared by the matrix in a SES system and those for SES-D (see Chiang, 1984, p. 614 and ss.). The behavior of the SES-D is already contained in the SES system in both their particular, or equilibrium, solution and their dynamic behavior.

We can now focus on the relationship between exact and approximate versions of SES. Again, note the following two facts:

1.  The particular, or equilibrium, solution of a SES is also the centered solution of the AES. This means that the AES solution encompasses the exact and equilibrium solutions of a dynamic SES, of course given the assumptions employed in Chapter 2 (Banach spaces, existence of a centered form, and so on).
2.  When a dynamic SES is convergent, the AES solution must—of necessity, given our assumptions—encompass the complementary solutions around equilibrium solutions. The reach of the approximate solution depends upon the size of the error employed when calculating the ranges of uncertainty. This means that for convergent systems, a succession of boxes (AES) could help us approximate the dynamic behavior when the exact dynamic paths and solutions are unavailable because we do not know enough to specify the dynamic behavior of the system. In economics and finance we are exclusively interested in convergent systems, because they carry equilibrium interpretations.

We have shown that AES can help in dynamic interpretation of static SES. It is appropriate to repeat some reflections about the modeling context in finance and economics, which reinforce our point:

1.  Important decisions are always forward-looking, and affect many periods ahead;
2.  The number of factors at play is unknown and even the known factors, and their relationships, cannot be quantified;
3.  The underlying processes are ever-changing and unlike physical or biological processes have no constants;
4.  We deal with humanistic systems whose behavior is not well known.

Given this context, and if anything better is not available, AES can perhaps help us to model and understand dynamic behavior.

## 4. AES AND PRACTICAL MODELING OPTIONS

It is appropriate to close this last chapter with a few words on modeling options alternative to AES.

Exhibit 6.5 classifies modeling options. When we reflect on the options available and what we normally see in published research and books, one cannot avoid being surprised at how little we experiment in economics and finance. In finance, especially, the range of models in the literature is appallingly narrow. This is most disappointing for several reasons. First, it deforms professionals, especially junior researchers, because the easy way to publish becomes adding small twists to existing models and by specializing in, for example, small distributions of some probabilistic specification. Second, it is a squandering of opportunities. Looking at the published literature, we would think the only way to model is to develop causal probability-calculus (CPC) applications, which could not be further from the truth. Moreover, the context in which modeling is to be developed in finance and economics—recall the four points at the end of the previous section—do not particularly favor CPC applications. The uncertainty is richer than that allowed by probabilistic specifications, and the intrinsic forward-looking character of the problems to be solved eliminates calculus from contention unless one uses approximations.

**Exhibit 6.5**
**A Sample of Modeling Options**

|  Information | Methodology |
| --- | --- |

*Symbolic*:

|  Qualitative | Algebra (Boolean, lattice, fuzzy sets-based, and so on), it is always causal. |
| --- | --- |

|  Quantitative | Causal: |
| --- | --- |

Calculus
Mathematical programming
Probabilistic
Noncausal, or connectionist, modeling: Neural
  networks (some neural nets are causal).

*Linguistic*:

Discourse analysis (inductive)
Propositional calculus (deductive)
Artificial intelligence
  Symbolic (schemas-based)
    (inductive)
  Procedural or rule based    (deductive)

The modeling context within finance and economics—if economics ever decides to adopt a "forward-looking" thrust—seems to favor nonprobabilistic, dynamic models perhaps in the form of approximate equations. It also favors qualitative and linguistic modeling, which are modeling options never to be seen in the current literature. Note that these are also forms of approximate modeling.

Independent of the specific approach adopted, we believe two main tasks will dominate modeling efforts in the near-medium term in economics and finance. The first is the search for significant structure in the form of reliable classes and categories that withstand change. This search will, in turn, cast light upon the building blocks and characteristics of the model. The second task will be to clarify the relationship between words and numbers in modeling, which we have not addressed yet, and which is needed to strengthen our choice between quantitative and qualitative models. Note that the area of confluence between words and numbers is also the confluence between mathematics (the algebra of sets) and philosophy (symbolic logic).

## CONCLUDING REMARKS (EPILOG)

As the hawk that saw an attractive prey many feet below, we developed a simple procedure to solve most common approximate equations systems, and then plunged into applying approximate equations to three areas that may immediately benefit from them.

SES are central to modern modeling. When we use AES, we do not have to abandon our A x = b model and develop a totally different one. Rather, we relax the phenomenal exactness of SES models, which limit their applicability. Moreover, as indicated in this chapter, SES models are the foundation of operational and classical dynamic models. A fact that is reinforced in a practical manner by reformulation of SES to approximate form.

Sometimes researchers fail to take advantage of what is available and too often pursue the most complicated option available. This book introduced a variety of SES that should present little technical difficulty to practitioners and researchers, but that appears, nonetheless, useful in practical applications. Perhaps this manual could be used as a stepping-stone into the more sophisticated developments of interval mathematics, which is now blooming. Note, however, that perhaps because of this emerging state, many of the works in interval mathematics seem to have been written by mathematicians for an audience of mathematicians, and some of the methods are very specialized. I feared these two characteristics would perhaps discourage the type of practitioners I precisely want to attract to approximate methods. The best practitioners in corporate finance and investments are eager to try new things but they want—and need—to understand what they do.

For the audience of researchers, two ideas from this chapter must be stressed. One, that practical research may require approximate and qualitative methods. Two, that we all work with representations we ourselves have made up, which also means that, at least in economics and finance, we all enjoy considerable freedom concerning modeling options.

For the audience of practitioners, it must be noted that the research in approximate methods presented in this book seems to suggest the following modeling sequence or research protocol:

1. First, try to study the data available, define variables and relationships, and develop a SES model for handling the problem. Contrary to what is often done in conventional modeling, one could favor streamlined, fundamental models with a handful of equations. These should express well-known relationships and well-defined variables. Avoid those procedures that add "fog" to the problem, for example, small distribution stochastics. Then study the numerical properties of the model and its ranges of uncertainty.

2.   Second, complement the approximate model with qualitative reasoning. Quantitative models may never be enough to support practical decision-making. This was certainly the case with macroeconomics. Even when approximate models appear very reliable, as in the case of financial planning and portfolio management, the most critical elements—strategic ones—are left out. Given the current state of knowledge in economics and finance, we still need natural language-based reasoning. (I feel that very little has been done in this area and am preparing to carry out an exploration of qualitative modeling for economics and finance).

3.   Third, explore other modeling approaches, such as neural nets and rules-based systems. There is no evidence that indicates these should be multimillion-dollar initiatives involving armies of scholars and expensive computing resources. I have the strong perception that it is the other way around: those models which business decision-makers may find most useful are likely to be fundamental and easy to understand and manage. Everything that is really useful in our lives often seems to exhibit those characteristics.

In any case, practical modeling in economics and finance simply illustrates what Tagore put best: "If you shut your door to all errors, truth will be shut out." When we do let errors come in, as in the case of AES, we find at the end of the road a black box that we have called the "ignorance box," and some others may have called Pandora's box. This should be expected. Given our limited intellectual and real resources, we will always find a wall at the end of our territories marking the beginning of yet another "terra incognita." We can push the wall further, but with the wall also goes our power to imagine what may lie beyond it; and that is perfectly all right with us. What would we do without having walls to push and without the unknown dangers and adventures they encircle?

I will finish this monograph by going back to Descartes, whose method paved the way for developments in SES, which constitute the grounding for this book. He said (see Caws, 1988, p. 163):

The sciences found in books do not approach so near the truth as the simple reasoning which a man of common sense can naturally carry out respecting the things that come immediately before him.

Our intention in writing this monograph was to provide a class of models that can be adapted to individual circumstances and practical settings. It should represent the best option to retain the benefits of centuries of modeling via SES without foregoing addressing some of its limitations. This, in sum, would permit one to blend the knowledge found in the sciences with the common sense found in practitioners.

# REFERENCES

Alchian, A. and Demsetz, H. Production, Information Costs and Economic Organization. *American Economic Review*, 1972, 777–795.

Alefeld, G., and Herzberger, J. *Introduction to Interval Computations*. Academic Press, 1983.

Alexander, G. Short-Selling and Efficient Sets. *The Journal of Finance*, Vol. XLVIII, No. 4, September 1993, p. 1497–1506.

———. Efficient Sets, Short Selling, and Estimation Risk. *Journal of Portfolio Management*, Winter 1995, 64–73.

Ansbacher, H., and Ansbacher, R., ed. *The Individual Psychology of Alfred Adler*. Harper Torchbooks, 1964.

Ayer, A. *Bertrand Russell*. The University of Chicago Press, 1972.

Balas, E. An Additive Algorithm for Solving Linear Programs with Zero–One Variables. *Operations Research,* 13, 1965, 517–546.

———. Discrete Programming by the Filter Method. *Operations Research,* 15, 1967, 915–957.

Barnett, S. *Matrices: Methods and Applications*. Clarendon Press, 1992.

Barro, R. *Macroeconomic Policy*. Harvard University Press, 1990.

Barry, C. Portfolio Analysis under Uncertain Means, Variances, and Covariances. *Journal of Finance*, Vol. 29, No. 2, 1974, 515–522.

Barry, C., and Winkler, R. Nonstationarity and Portfolio Choice. *Journal of Financial and Quantitative Analysis*, Vol. XI, No. 2, June 1976, 217–235.

Beckenback, E., and Bellman, R. *An Introduction to Inequalities*. Random House, 1961.

Begg, D. *The Rational Expectations Revolution in Macroeconomics: Theories and Evidence*. Johns Hopkins University Press, 1982.

Berlinski, D. *A Tour of the Calculus*. Vintage Books, 1995.

Black, F. The Magic in Earnings: Economic Earnings Versus Accounting Earnings. *Financial Analyst Journal*, November-December 1980, 19–24.

———. The Trouble with Econometric Models. *Financial Analyst Journal*, March-April, 1982, 29–37.

Bodie, Z., and Merton, R. *Finance*. Prentice-Hall, 1998.

Bogle, J. Selecting Equity Mutual Funds. *Journal of Portfolio Management*, Winter 1992, 94–100.

———. *Bogle on Mutual Funds: New Perspectives for the Intelligent Investor.* Irwin, 1994.

Bolles, R. *The History of Psychology: A Thematic History.* Brooks/Colle Publishing Co., 1993.

Bookstaber, R. *Option Pricing and Investment Strategies.* Probus Publishing, 1989.

Bookstaber, R., and Clarke, R. Options Can Alter Portfolio Return Distributions. The *Journal of Portfolio Management*, Vol. 7, 1981, 63–70.

Boot, J. *Quadratic Programming.* North-Holland, 1964.

Brennan, M. The Optimal Number of Securities in a Risky Asset Portfolio When There Are Fixed Costs of Transacting: Theory and Some Empirical Results. *Journal of Financial and Quantitative Analysis*, Vol. 10, No. 3, September 1975, 483–496.

Brigham, E., and Gapenski, L. *Intermediate Financial Management.* 4th ed. Dryden Press, 1993.

Bronshtein, I., and Semendyayev, K. *Handbook of Mathematics.* Van Nostrand Reinhold Company, 1985.

Burrill, C., and Quinto, L. *Computer Model of a Growth Company.* Gordon and Breach Science Publishers, 1972.

Buser, S. Mean-Variance Portfolio Selection with either a Singular or Non-singular Variance-Covariance Matrix. *Journal of Financial and Quantitative Analysis*, Vol. XII, No. 3, September 1977, 436–461.

Campbell, J., Lo, A., and MacKinlay, C. *The Econometrics of Financial Markets.* Princeton University Press, 1997.

Carleton, W. An Analytical Model for Long Term Financing. *Journal of Finance*, Vol. 25, No. 2, May 1970, 291–315.

Carleton, W., Dick, J., and Downes, D. Financial Policy Models: Theory and Practice. *Journal of Financial and Quantitative Analysis*, Vol. 8, December 1973, 691–709. Also in Myers ed. *Modern Developments in Financial Management.* Praeger Publishers, 1976, 564–582.

Carter, R., and Auken, H. Security Analysis and Portfolio Management: A Survey and Analysis. *Journal of Portfolio Management*, Vol. 16, No. 3, 1990, 81–85.

Caws, Peter. *Structuralism for the Human Sciences.* Humanities Press, 1988, reprinted 1997.

Chance, D. *An Introduction to Derivatives.* 3rd ed. Dryden, 1995.

Chiang, A. *Fundamental Methods of Mathematical Economics.* 3rd ed. McGraw-Hill, 1984.

Chow, G., and Corsi, P. *Macroeconomic Models.* John Wiley and Sons, 1982.

Cohon, J. *Multiobjective Programming and Planning.* Academic Press, 1978.

Connor, G. A Unified Beta Pricing Theory. *Journal of Economic Theory*, 34, 1984, 13–31.

Copeland, J., and Weston, F. *Financial Theory and Corporate Policy.* Addison Wesley, 1988.

Corman, J., and Lussier, R. *Small Business Management: A Planning Approach.* Irwin, 1996.

Courant, R, Robbins, H., and Stewart, I. *What Is Mathematics?* Oxford University Press, 1996.

Coveney, P., and Highfield, R. *Frontiers of Complexity: The Search for Order in a Chaotic World.* Fawcett Coumbine, 1995.

Cox, J., Ingersoll, J., and Ross, S. An Intertemporal General Equilibrium Model of Asset Prices. *Econometrica*, 53, 1985, 363–384.

Cox, J., and Rubinstein, M. *Option Markets*. Prentice-Hall, 1985.

Curley, M. *Understanding and Using Margin*. Probus Publishing Company, 1989.

Deif, A. *Sensitivity Analysis in Linear Systems*. Springer-Verlag, 1986.

Deleuze, G., and Guattari, F. *What Is Philosophy?* Columbia University Press, 1994.

Descartes, R. *Discourse on Method and the Meditations*. Penguin Classics, 1968.

Dickinson, J. The Reliability of Estimation Procedures on Portfolio Analysis. *Journal of Financial and Quantitative Analysis*. Vol. 9, No. 3, 1974, 447–463.

Dothan, M. *Prices in Financial Markets*. Oxford University Press, 1990.

Downing, D., and Clark, J. *Quantitative Methods*. Barron's Business Review, 1988.

Drazin, P. *Nonlinear Systems*. Cambridge University Press, 1992.

Dubois, D., and Prade, H. *Fuzzy Sets and Systems*. Academic Press, 1980.

Duffie, D. *Security Markets: Stochastic Models*. Academic Press, 1988.

———. *Dynamic Asset Pricing Theory*. Princeton University Press, 1992.

Elton, E., and Gruber, M. Risk Reduction and Portfolio Size: An Analytical Solution. *Journal of Business*, No. 4, October 1977, 415–437.

———. *Modern Portfolio Theory and Investment Analysis*. 5th ed. Wiley, 1995.

Ewald, G. *Geometry: An Introduction*. Wadsworth Publishing, 1971.

Fabozzi, F. *Bond Markets, Analysis and Strategies*. Prentice-Hall, 1993.

Fair, R. *Specification, Estimation, and Analysis of Macroeconometric Models*. Harvard University Press, 1984.

Fama, E. and Miller, M. *The Theory of Finance*. Holt, Rinehart and Winston, 1972.

Fiedler, E. What's New in Economics? *Across the Board*, May 1984, 49–54.

Fogler, R. A Modern Theory of Security Analysis. *Journal of Portfolio Management*, Spring 1993, 6–14.

Foucault, M. *The Order of Things: The Archeology of Human Science*. Vintage Books, 1994.

Foulds, L. Optimization Techniques. Springer-Verlag, 1981.

Francis, J. *Investments: Analysis and Management*. 4th ed. McGraw-Hill, 1986.

Francis, J., and Archer, S. *Portfolio Analysis*. Prentice-Hall, 1971.

Francis, J., and Rowell, D. R. A Simultaneous Equation Model of the Firm for Financial Analysis and Planning. *Financial Management*, Vol. 7, Spring 1978, 29–44.

Frankfurter, G., and Phillips, H. Measuring Risk and Expectation Bias in Well Diversified Portfolios. *TIMS Studies in Management Sciences*, Vol. 11, 1979, 73–77.

Frankfurter, G., Phillips, H., and Seagle, J. Portfolio Selection: The Effects of Uncertain Means, Variances and Covariances. *Journal of Financial and Quantitative Analysis*, Vol. 6, No. 5, 1971, 1251–1262.

Franklin, J. *Methods of Mathematical Economics: Linear and Non-linear Programming, Fixed Point Theorems*. Springer-Verlag, 1980.

Freund, J. *Mathematical Statistics*. Prentice-Hall, 1992.

From, G., and Klein, L., eds. *The Brooking Model: Perspectives and Development*. North Holland, 1975.

Frost, P., and Savarino, J. For Better Performance: Constraint Portfolio Weights. *Journal of Portfolio Management*, Vol. 15, No. 1, Fall 1988, 29–34.

Fry, F. *Entrepreneurship: A Planning Approach*. West Publishing Co. 1992.

Gal, T. *Post-optimal Analyses, Parametric Programming, and Related Topics*. McGraw-Hill, 1979.

Garbade, K. *Securities Markets*. McGraw-Hill, 1982

Garfinkel, R., and Nemhauser, G. *Integer Programming*. Wiley, 1972.

Genotte, G., and Jung, A. Commissions and Asset Allocation. *Journal of Portfolio Management*, Fall 1992, 12–23.

Geoffrion, G. Linear Programming by Implicit Enumeration and Balas' Method. *SIAM Review* 9, 1967, 178–190.

Gitman, L. *Principles of Corporate Finance*. HarperCollins, 1993.

Gleick, J. *Chaos*. Penguin Books, 1987.

Glover, F. A Note on Integer Programming and Integer Feasibility. *Operations Research* 16, 1968, 1212–1216.

Hammer, R., Hocks, M., Kulisch, U., and Ratz, D. *C++ Toolbox for Verified Computing*. Springer-Verlag, 1995.

Hansen, E. *Global Optimization Using Interval Analysis*. Marcel Dekker, 1992.

Hansen, E. Interval Arithmetics in Matrix Computations. Part I. *SIAM Journal of Numerical Analysis*, 2, 1965, 308–320.

Hansen, E., and Smith, R. Interval Arithmetics in Matrix Computations. Part II. *SIAM Journal of Numerical Analysis*, 4, 1967, 1–9.

Harrington, D. *Modern Portfolio Theory, The Capital Asset Pricing Model & Arbitrage Pricing Theory: A User's Guide*. 2nd ed. Prentice-Hall, 1987.

Haugen, R. *The New Finance: The Case Against Efficient Markets*. 2nd ed. Prentice-Hall, 1995.

———. Finance from a New Perspective. *Financial Management*, Vol. 25, No. 1, Spring 1996, 86–87.

———. *Modern Investment Theory*. 4th ed. Prentice-Hall, 1997.

———. *Beast on Wall Street: How Stock Volatility Devours Our Wealth*. Prentice-Hall, 1999a.

———. *The Inefficient Stock Market: What Pays off and Why*. Prentice-Hall, 1999b.

Henderson, J., and Quandt, R. *Microeconomic Theory: A Mathematical Approach*. McGraw-Hill Book Company, 1980.

Higgins, R. How Much Growth Can a Firm Afford? *Financial Management*, Vol. 6, Fall 1977, 7–16.

———. *Analysis for Financial Management*. Irwin, 1995.

Hsia, C. Coherence of the Modern Theories of Finance. *Financial Review*, Vol. 16, No. 1, 1981, 27–42.

Huang, C., and Litzenberger, R. *Foundations for Financial Economics*. Prentice-Hall, 1988.

Jacob, N. A Limited-Diversification Portfolio Selection Model for the Small Investor. *Journal of Finance*, Vol. 29, No. 3, 1974, 847–856.

Jarrow, R., and Madan, D. Is Mean-variance Analysis Vacuous: Or Was Beta Stillborn? *European Finance Review*, 1, 1997, 15–30.

Jarrow, R., and Turnbull, S. *Derivative Securities*. South-Western, 1996.

Jobson, J., and Korkie, B. Estimation for Markowitz Portfolios. *Journal of the American Statistical Association*, Vol. 75, Number 371, September 1980, 544–554.

———. Putting Markowitz Theory to Work. *The Journal of Portfolio Management*, Summer, Vol. 7, No. 4, 1981, 70–74.

Johnston, J. *Econometric Methods*. 3rd ed. McGraw-Hill, 1984.

Judge, G., Griffiths, W., Carter Hill, R., Lutkepohl, H., and Lee, T. *The Theory and Practice of Econometrics*. 2nd ed. Wiley, 1985.

Kantor, B. Rational Expectations and Economic Thought. *Journal of Economic Literature*, Vol. XXVII, December 1979, 1242–1441.

Klein, L. *Economic Fluctuations in the United States: 1921–1941*. John Wiley and Sons, 1950.

Klein, L., and Young, R. *An Introduction to Econometric Forecasting and Forecasting Models*. Lexington Books, 1980.

Kline, M. *Mathematics for the Non-Mathematician*, Dover, 1967.

Knight, W. Working Capital Management—Satisficing versus Optimization. *Financial Management*, 1, Spring 1972, 33–40.

Kolmogorov, A, and Fomin, S. *Introductory Real Analysis.* Dover Publications, 1970.

Kramer, E. *The Nature and Growth of Modern Mathematics.* Princeton University Press, 1981.

Kreinovich, V., Lakeyev, A., and Noskov, S. Optimal Solution of Interval Linear Systems is Intractable (NP-hard). *Interval Computations*, No. 1, 1993, 6.

Kritzman, M. What Practitioners Need to Know about Optimization. *Financial Analyst Journal*, 48, September-October 1992, 10–18.

Kuperman, I. *Approximate Linear Algebraic Equations.* Van Nostrand Reinhold Company, 1971.

Kuttner, R. The Poverty of Economics. *The Atlantic Monthly*, February 1985, 74–84.

Kwan, C. Portfolio Analysis Using Single Index, Multi-index, and Constant Correlation Models: A Unified Treatment. *The Journal of Finance*, IXL, June 1984, 1469–1483.

Lai, Y., and Hwang, C. *Fuzzy Mathematical Programming and Applications.* Springer-Verlag, 1993.

Lakonishok, J., Schleifer, J., and Vishny, R. The Structure and Performance of the Money Management Industry. *Brookings Papers on Economic Activity, Microeconomics*, 1992, 339–391.

Langer. *Symbolic Logic.* Dover, 1967.

Lease, R., Lewellen, W., and Schlarbaum, G. Market Segmentation: Evidence on the Individual Investor. *Financial Analyst Journal*, No. 5, September-October 1976, 53–60.

Lee, C. *Financial Analysis and Planning: Theory and Application.* Addison-Wesley, 1985.

Lee, C., and Finnerty, J. *Corporate Finance. Theory, Methods and Applications.* Harcourt, Brace and Jovanovich, 1990.

Lee, W. Diversification and Time: Do Investment Horizons Matter? *The Journal of Portfolio Management*, Vol. 16, No. 3, Spring 1990, 21–26.

Leibowitz, M., and Henriksson, R. Portfolio Optimization with Short-fall Constraints: A Confidence Limit Approach to Managing Portfolio Risk. *Financial Analyst Journal*, Vol. 45, No. 2, Fall 1989, 31–41.

Levy, H., and Sarnat, M. *Portfolio and Investment Selection: Theory and Practice.* Prentice-Hall, 1984.

Levy, H., and Yoder, J. Applying the Black-Scholes Model after Large Market Schocks. *Journal of Portfolio Management*, Vol. 16, No. 1, 1989, 103–106.

Lintner, J. The Valuation of Risk Assets and the Selection of Risky Investments in Stock Portfolios and Capital Budgets. *Review of Economics and Statistics*, 47, 1965, 13–37.

Lipschutz, S. *300 Solved Problems in Linear Algebra.* Schaum's Outline Series, McGraw-Hill, 1989.

———. *Linear Algebra.* 2nd ed. Schaum's Outline Series, McGraw-Hill, 1991.

Lipschutz, S., and Lipson, M. *Discrete Mathematics.* Schaum's Outline Series, McGraw-Hill, 1997.

Luenberger, D. *Linear and Non-Linear Programming.* 2nd ed. Addison-Wesley, 1984.

———. *Investment Science.* Oxford University Press, 1998.

Luskin, D. *Index Options and Futures: The Complete Guide.* Wiley, 1987.

MacMillan, L. *Options as a Strategic Investment.* New York Institute of Finance, 1986.

Maddala, G. *Econometrics.* McGraw-Hill, 1977.

Madrick, J. A New Economics on Wall Street. *New York Times*, Sunday, March 6, 1988.

Malmgren, H. B. Information, Expectations and the Theory of Firms. *Quarterly Journal of Economics*, 75, August 1961, 399–421.

Markowitz, H. Portfolio Selection. *The Journal of Finance*, Vol. 12, March 1952, 77–91.

———. *Portfolio Selection, Efficient Diversification of Investments*. New York, John Wiley and Sons, 1959.

———. Nonnegative and Not Nonnegative: A Question about CAPMs. *The Journal of Finance*, Vol. XXXVIII, No. 2, May 1983, 283–295.

———. Foundations of Portfolio Theory. *The Journal of Finance*, Vol. XLVI, No. 2, June 1991, 469–477.

Martin, A. Mathematical Programming of Portfolio Selections. *Management Science*, Vol. 1, No. 2, January 1955, 152–166. Reprinted in E. B. Frederickson, *Frontiers of Investment Analysis*. International Textbook, 1965, 367–381.

Merton, R. An Analytic Derivation of the Efficient Frontier. *Journal of Financial and Quantitative Analysis*, Vol. VII, No. 4, September 1972, 1851–1872.

———. On Estimating the Expected Return on the Market. *Journal of Financial Economics*, 8, 1980, 323–361.

Michaud, R. The Markowitz Optimization Enigma: Is 'Optimized' Optimal? The *Financial Analyst Journal*, January-February 1989, 31–42.

———. *Efficient Asset Management*. Harvard Business School Press, 1998.

Miller, E. Divergence of Opinion, Short Selling, and the Role of the Marginal Investor. In Frank Fabozzi, ed. *Managing Institutional Assets*. Harper and Row Publishers, 1990, 143–183.

———. Arbitrage Pricing Theory: A Graphical Critique. Non-linear Relationships Are Possible. *The Journal of Portfolio Management*, Vol. 18, Fall 1991, 72–76.

Moore, R. *Interval Analysis*. Prentice-Hall Series in Automatic Computation, 1966.

———. Methods and Applications of Interval Analysis. *Siam Studies in Applied Mathematics*, 1979.

Morishima, M., Murata, Y., Noose, T., and Saito, M. *The Working of Econometric Models*. Cambridge University Press, 1972.

Mossin, J. Equilibrium in a Capital Asset Market. *Econometrica*, 34, 261–76.

Naylor, T. *Computer Simulation Experiments with Models of Economic Systems*. John Wiley, 1971.

Nemhauser, G. and Wolsey, L. *Integer and Combinatorial Optimization*. Wiley, 1988.

Neumaier, A. *Interval Methods for Systems of Equations*. Cambridge University Press, 1990.

Newbould, G., and Poon, P. The Minimum Number of Stocks Needed for Diversification. *Financial Practice and Education*, Vol. 3, No. 2, Fall 1993, 85–87.

Oettli, W. On the Solution Set of a Linear System with Inaccurate Coefficients. *SIAM Journal of Numerical Analysis*, Series B, Vol. 2, No. 1, 1965, 115–118.

Oettli, W., Prager, W., and Wilkinson, J. Admissible Solutions of Linear Systems with Not Sharply Defined Coefficients. *SIAM Journal of Numerical Analysis*, Series B, Vol. 2, No. 2, 1965, 291–299.

Pettofrezzo, A. *Matrices and Transformations*. Dover, 1966.

Petty, W., Keown, A., Scott, D., Martin, J. and Petite. Financial Management: A Small-Firm Perspective. Chapter 26 in *Basic Financial Management*. Prentice-Hall, 1993.

Pierre, D. *Optimization Theory with Applications*. Dover, 1986.

Poole, W. Optimal Choice of Monetary Policy Instruments in a Simple Stochastic Macromodel. *Quarterly Journal of Economics*, May 1970, 197–216.

Posner, M., ed. *Foundations of Cognitive Science*. MIT Press, 1993.

Radcliffe, R. *Investments: Concepts, Analysis and Strategies*. 5th ed. HarperCollins, 1997.

Ritchken, P. *Options: Theory, Strategy, and Applications*. Scott, Foreman and Company (HarperCollins), 1987.

Rockafellar, T. *Convex Analysis*. Princeton University Press, 1970.

Roll, R., and Ross, S. A Critical Reexamination of the Empirical Evidence of the Arbitrage Pricing Theory: A Reply. *The Journal of Finance*, Vol. XXIX, No. 2, June 1984, 347–350.

Rorty, M. *Philosophy and the Mirror of Nature*. Princeton University Press, 1979.

Ross, S. The Arbitrage Theory of Capital Asset Pricing. *Journal of Economic Theory*, 13, 1976, 341–360.

———. Risk, Return and Arbitrage. In Friend, I., and Bickster, J. *Risk and Return in Finance*, Ballinger, 1977.

Russell, B. *The Philosophy of Logical Atomism*. Open Court Classics, 1985.

Samuelson, P. *Foundations of Economic Analysis*. Enlarged edition. Harvard University Press, 1983.

Sargent, T. *Macroeconomic Theory*. Academic Press, 1987.

Schopenhauer, A. *The World as Will and as Representation*. Two volumes. Dover, 1958.

Sears, S., and Trennepohl, G. *Investment Management*. The Dryden Press, 1993.

Shackle, G. *Epistemics and Economics*. Cambridge University Press, 1972.

Sharpe, W. A Simplified Model for Portfolio Analysis. *Management Science*, 9, 1963, 277–293.

———. Capital Asset Prices: A Theory of Market Equilibrium under Conditions of Risk. *The Journal of Finance*, Vol. XIX, No. 3, September 1964, 425–442.

———. A Linear Programming Algorithm for Mutual Fund Portfolio Selection. *Management Science*, Vol. 13, No. 7, March 1967, 499–510.

———. Capital Asset Prices with and without Negative Holdings. *The Journal of Finance*, Vol. XLVI, No. 2, June 1991, 489–509.

Shilov, G. *Linear Algebra*. Dover, 1977.

Smith, K. Alternative Procedures for Revising Investment Portfolios. *Journal of Financial and Quantitative Analysis*, Vol. III, No. 4, December. 1968, 371–403.

———. The Major Asset Mix Problem of the Individual Investor. *Journal of Contemporary Business*, Winter 1974. Also in Keith Smith ed. *Case Problems and Readings: A Supplement for Investments and Portfolio Management*. Reading 16, p. 279. McGraw-Hill, 1990.

Spiegel, M. *Statistics*. 2nd ed. Schaum's Outline Series, McGraw-Hill, 1992.

Spielberg, K. Enumerative Methods in Integer Programming. *Annals of Discrete Mathematics*, 5, 1979, 139–183.

Spivey, W. Econometric Model Performance in Forecasting and Policy Assessment. *American Enterprise Institute for Public Policy Research*, 1979. AEI Studies: 225.

Statman, M. How Many Stocks Make a Diversified Portfolio? *Journal of Financial and Quantitative Analysis*, Vol. 22, 1987, No. 3, 353–364.

Strogatz, S. *Nonlinear Dynamics and Chaos*. Addison Wesley, 1994.

Tarrazo, M. (1990). Direct Investment Theory for the Individual (Small) Investor: The State of the Art. 1990 Meeting of the Academy of Financial Services. Orlando, Florida, October.

———. (1992). Limited Diversification, Common Stock Options and Individual Investors. 1992 Meeting of the Academy of Financial Services. San Francisco, October 21.

———. (1997a). An Application of Fuzzy Set Theory to the Individual Investor Problem. *Financial Services Review*, Vol. 6, No. 2, 1997, 97–107.

————. (1997b). Is Finance a Paper Tiger? A Critical Analysis of the Methodology of Finance. Proceedings of the 1997 Meeting of the Academy of Economics and Finance. Lafayette, Louisiana, February 12–15.

————. (1997c). On The Application of Interval Mathematics and Fuzzy Logic to Economics and Business Decision-Making: A Conceptual Analysis. 1997 Meeting of the Midwest Finance Association. Kansas City, Missouri, March 20–22.

————. (1997d). Interval Mathematics and Fuzzy Mathematical Programming. Seminar offered by the author at the 1997 Meeting of the Midwest Finance Association. Kansas City, Missouri, March 20–22.

————. (1997e). Quadratic Forms of Binary Variables: An Application to Portfolio Optimization. 1997 Meeting of the Western Decision Sciences Institute. Big Island of Hawaii, March 25–29.

————. (1997f). Post-Optimal Analysis of Investment Portfolios and Basic Indeterminacy of Portfolio Weights. 1997 Meeting of the Eastern Finance Association. Panama City Beach, Florida, April 19.

————. (1997g). A General Framework for the Analysis of Economic and Business Decisions. 1997 Meeting of Western Social Science Association, Albuquerque, New Mexico, April 24.

————. (1997h). Rethinking Modeling in the Information Age. 1997 Meeting of Western Social Science Association, Albuquerque, New Mexico, April 24.

————. (1997i). A Methodology and Model for Qualitative Business Planning. *International Journal of Business Research*, Vol. 3, No. 1, Fall 1997, 41–62.

————. (1998a) Calculating Uncertainty Intervals in Approximate Equations Systems. *Applied Numerical Mathematics,* 26, 1998, 1–9.

————. (1998b) More on the 5% Rule. *Midwest Review of Finance and Insurance*, Vol. 12, No. 1, March 1998, 344–353.

————. (1999) Fuzzy Sets and the Investment Decision. *Financial Technology*, June 1999, pp. 37–47, Institutional Investor Journals.

————. (2000a). Trading Restrictions and Revision Programs of Optimal Portfolios. 29th Meeting of the Western Decision Sciences Institute. April 18–22, 2000. Island of Maui, Hawaii.

————. (2000b) Mathematical Programming and Portfolio Optimizations: A Clarification. *Advances in Financial Planning and Forecasting,* Vol. 9, 31–55.

Tarrazo, M., and Alves, G. *Advanced Spreadsheet Modeling for Portfolio Management.* Kendall-Hunt, 1998 (several reprints). (1998a).

————. (1988b) Portfolio Optimization Under Realistic Short Sales Conditions. *International Journal of Business*, Vol. 3, No. 2, 1998, 77–93.

Tarrazo, M., and Gutierrez, L. (1997) Perspectives on Financial Planning. *Research Papers in Management and Business*, 1997, 61–79. Ecole Superieure de Commerce, Montpellier-ESKA Editions, Paris, France.

————. (2000) Economic Expectations, Fuzzy Sets, and Financial Planning. *European Journal of Operational Research*, Vol. 126, No. 1, August 2000, 89–105.

Theil, H., and Van de Panne, C. Quadratic Programming as an Extension of Classical Quadratic Maximization. *Management Science*, Vol. 7, No. 1, October 1960, 1–20.

Thurow, L. *Dangerous Currents in Economics*. Vintage, 1984.

Tobin, J. Liquidity Preference as Behavior Towards Risk. *Review of Economic Studies* 25, 1958, 65–86.

Tucker, A., Becker, K., Isimbabi, M., and Odgen, J. *Contemporary Portfolio and Risk Management*. West Publishing, 1994.

UN-ECE (United Nations. Economic Commission for Europe). Macroeconomic Models for Planning and Policy-making. Edited by the Secretariat of the Economic Commission for Europe. Geneva, 1967. E/ECE/665. United Nations publications. Sales no.: E.67.II.E.3.

Varian, H. *Microeconomic Analysis*. Norton, 1984.

Wagner, W., and Banks, M. Increasing Portfolio Effectiveness Via Transaction Costs. *The Journal of Portfolio Management*, Fall 1992, 6–11.

Warren, J., and Shelton, J. A Simultaneous Equation Approach to Financial Planning. *Journal of Finance*, Vol. XXVI, December 1971, 1123–1142.

Wolfe, P. The Simplex Method for Quadratic Programming. *Econometrica*, 27, 1959, 382–398.

Zadeh, L. Outline of a New Approach to the Analysis of Complex Systems and Decision Processes. *IEEE Transactions on Systems, Man, and Cybernetics*, 3, 1973, 28–44.

# INDEX

**About the Author**

MANUEL TARRAZO is Professor of Finance at the University of San Francisco. He is the author of several scholarly articles and coauthor of a previous book.